THE COMMUNITY DEVELOPMENT
QUOTA PROGRAM IN
ALASKA
AND LESSONS FOR THE
WESTERN PACIFIC

Committee to Review the Community Development Quota Program

Ocean Studies Board

Commission on Geosciences, Environment, and Resources

National Research Council

NATIONAL ACADEMY PRESS
Washington, DC 1999

This report and the committee were supported by the National Oceanic and Atmospheric Administration. The views expressed herein are those of the authors and do not necessarily reflect the views of the sponsor.

This study was supported by Contract No. 50-DKNA-6-90040, Task No. 7-95103 between the National Academy of Sciences and the National Oceanic and Atmospheric Administration. Any opinions, findings, conclusions, and recommendations expressed in this publication are those of the author(s) and do not necessarily reflect the view of the organizations or agencies that provided support for the project.

Cover: Double hoop halibut spirit mask (approximately 24 inches tall) made of driftwood by Yupik artist Lola Ferguson, Nunivak Island, Alaska. Used with permission of the artist; photo provided by the Department of the Interior, Indian Craft Shop, Washington, D.C.

Library of Congress Catalog Card Number 98-86773
International Standard Book Number 0-309-06082-6

Additional copies of this report are available from:

National Academy Press
2101 Constitution Avenue, NW
Box 285
Washington, DC 20055
(800) 624-6242
(202) 334-3313 (in the Washington Metropolitan Area)
http://www.nap.edu

iv

The National Academy of Sciences is a private, nonprofit, self-perpetuating society of distinguished scholars engaged in scientific and engineering research, dedicated to the furtherance of science and technology and to their use for the general welfare. Upon the authority of the charter granted to it by the Congress in 1863, the Academy has a mandate that requires it to advise the federal government on scientific and technical matters. Dr. Bruce Alberts is president of the National Academy of Sciences.

The National Academy of Engineering was established in 1964, under the charter of the National Academy of Sciences, as a parallel organization of outstanding engineers. It is autonomous in its administration and in the selection of its members, sharing with the National Academy of Sciences the responsibility of advising the federal government. The National Academy of Engineering also sponsors engineering programs aimed at meeting national needs, encourages education and research, and recognizes the superior achievements of engineers. Dr. William A. Wulf is president of the National Academy of Engineering.

The Institute of Medicine was established in 1970 by the National Academy of Sciences to secure the services of eminent members of appropriate professions in the examination of policy matters pertaining to the health of the public. The Institute acts under the responsibility given to the National Academy of Sciences by its congressional charter to be an adviser to the federal government and, upon its own initiative, to identify issues of medical care, research, and education. Dr. Kenneth I. Shine is president of the Institute of Medicine.

The National Research Council was organized by the National Academy of Sciences in 1916 to associate the broad community of science and technology with the Academy's purposes of furthering knowledge and advising the federal government. Functioning in accordance with general policies determined by the Academy, the Council has become the principal operating agency of both the National Academy of Sciences and the National Academy of Engineering in providing services to the government, the public, and the scientific and engineering communities. The Council is administered jointly by both Academies and the Institute of Medicine. Dr. Bruce M. Alberts and Dr. William A. Wulf are chairman and vice chairman, respectively, of the National Research Council.

Preface

The Community Development Quota (CDQ) program was designed to improve social and economic conditions in rural western Alaska by helping communities build their capacity to engage in commercial fishing. Like all new efforts, the program has had some start-up difficulties. But as indicated in this review conducted by the Committee to Review the Community Development Quota Program, there has been significant progress and the program offers a great deal of promise for this particular region. Whether a similar program might be effective in other parts of Alaska or in the western Pacific, where there has been interest in the approach, is less clear because of differences in the environments, fishery management strategies, and the nature of the communities.

As chair of this committee, and as a member of the Ocean Studies Board, I would like to thank my fellow committee members for their hard work, patience, and cooperation. They were drawn from diverse fields, yet quickly and efficiently dealt with the complexities of the CDQ program. Moreover, they made a special effort to visit many communities involved with the CDQ program and, as a result, they obtained a variety of perspectives and experiences. The many people who contributed are thanked in Appendix F and the sites visited are listed in Chapter 1. We offer special thanks to the following people for their diligence in providing information for the committee's use: Clarence Pautzke and the rest of the staff at the North Pacific Fishery Management Council in Anchorage, Alaska; Kitty Simonds and the staff at the Western Pacific Fishery Management Council in Honolulu, Hawaii; the many NOAA staff who helped (including William Fox, Amy Gautam, Jay Ginter, Sally Bibb, Sam Pooley, and Ray Clarke); Glenn

Haight, Department of Community and Regional Affairs in Juneau, Alaska; and Julie Anderson, formerly with the Department of Community and Regional Affairs in Juneau, Alaska.

Our sincere appreciation goes to Elizabeth Clarke, who served as the study's director until she returned to her home institution, the University of Miami; and Chris Elfring, Director of the Polar Research Board, who saw the study through to completion. Each provided important leadership. We would also like to thank research associate Glenn Merrill and project assistant Ann Carlisle for their hard work on this project.

This report has been reviewed by individuals chosen for their diverse perspectives and technical expertise, in accordance with procedures approved by the NRC's Report Review Committee. This independent review provided candid and critical comments that assisted the authors and the NRC in making the published report as sound as possible and ensured that the report meets institutional standards for objectivity, evidence, and responsiveness to the study charge. The content of the review comments and draft manuscript remain confidential to protect the integrity of the deliberative process. We wish to thank the following individuals for their participation in the review of this report: Dr. Matt Berman, University of Alaska, Anchorage; Dr. Paul Callaghan, University of Guam, Mangilao; Dr. Nicholas Flanders, Institute of Arctic Studies, Hanover, New Hampshire; Mr. Zeke Grader, Pacific Coast Federation Fishermen's Associations, San Francisco, California; Dr. Patrick V. Kirch, University of California, Berkeley; Dr. Bonnie McCay, Rutgers University, Cook College, New Brunswick, New Jersey; Dr. H. Ronald Pulliam, University of Georgia, Athens; and Dr. Terrance Quinn, University of Alaska, Juneau.

While the individuals listed above provided many constructive comments and suggestions, responsibility for the final content of this report rests solely with the authoring committee and the NRC.

This study was requested by Congress as part of the Magnuson-Stevens Act of 1996. The request had two parts, this effort to review Community Development Quotas and another focused on Individual Fishing Quotas. Both reports are part of the Ocean Studies Board's continued effort to provide advice to Congress and the National Marine Fisheries Service on important fisheries issues.

John E. Hobbie
Chairman

Contents

THE COMMUNITY DEVELOPMENT QUOTA
PROGRAM IN
ALASKA

Executive Summary

The Community Development Quota (CDQ) program was implemented in December 1992 by the North Pacific Fishery Management Council. The CDQ program allocates a portion of the annual fish harvest of certain commercial species directly to coalitions of villages, which because of geographic isolation and dependence on subsistence lifestyles have had limited economic opportunities. The program is an innovative attempt to accomplish community development in rural coastal communities in western Alaska, and in many ways it appears to be succeeding. The CDQ program has fostered greater involvement of the residents of western Alaska in the fishing industry and has brought both economic and social benefits. The program is not without its problems, but most can be attributed to the newness of the program and the inexperience of participants. Overall the program appears on track to accomplishing the goals set out in the authorizing legislation: to provide the participating communities with the means to develop ongoing commercial fishing activities, create employment opportunities, attract capital, develop infrastructure, and generally promote positive social and economic conditions.

STRENGTHS AND WEAKNESSES OF THE CDQ PROGRAM

Because the program is still relatively new, the data necessary for detailed evaluation are limited and it is not yet possible to detect long-term trends. The six CDQ groups, organized from the 56 eligible communities (later expanded to 57), were of varying sizes and took varying approaches to harvesting their quota and allocating the returns generated. Although not all groups have been equally suc-

1

cessful, there were significant examples of real benefits accruing to the communities. All six groups saw creation of jobs as an important goal and stressed employment of local residents on the catcher-processor vessels and shoreside processing plants. All incorporated some kind of education and training component for residents, although to different degrees and with different emphases. Another benefit of the program is that the periodic nature of employment in the fishing industry preserves options for the local people to continue some elements of their subsistence lifestyles. The CDQ program generates resources that give local communities greater control of their futures. The State of Alaska also has played its part relatively effectively—it was efficient in reviewing the Community Development Plans, monitoring how the communities progressed, and responding to problems. Some of these responses, like reallocating quota share among communities, have been controversial, as might be expected.

Perhaps the greatest weakness of the CDQ program as implemented is a lack of open, consistent communication between the CDQ groups and the communities they represent, particularly a lack of mechanisms for substantial input from the communities into the governance structures. There has also been a lack of outreach by the state to the communities to help ensure that the communities and their residents are aware of the program and how to participate. For the CDQ program to be effective there must be a clear, well-established governance structure that fosters exchange of information among the groups' decisionmakers, the communities they represent, and the state and federal personnel involved in program oversight.

Some debate has centered on uncertainty about the intended beneficiaries of the program. It is unclear whether the program is intended primarily for the Native Alaskan residents of the participating communities or, if not, whether the governance structures should be modified to ensure that non-Native participation is possible. Similarly, there has been dissatisfaction among segments of the fishing industry that are not involved, either directly or as partners of CDQ groups, who believe that the program unfairly targets a particular population for benefits. This conflict is inevitable, given that the CDQ program is designed to provide opportunities for economic and social growth specifically to rural western Alaska. This policy choice specifically defines those to be included and cannot help but exclude others.

Although it is logical to require initially that all reinvestment of profits be in fishery-related activities because the initial objective of the CDQ program is to help the participating communities to establish a viable presence in this capital intensive industry, over time there should be more flexibility in the rules governing allocation of benefits—perhaps still requiring most benefits to be reinvested in fishing and fisheries-related activities but allowing some portion to go to other community development activities. This will better suit the long-term goal of the program, which is development of opportunities for communities in western Alaska.

The main goal of the CDQ program—community development—is by definition a long-term goal. Thus there is a need for a set and dependable program duration and the certainty that brings to oversight and management. This will allow CDQ group decisionmakers to develop sound business plans and will reduce pressures to seek only short-term results. However, calling for the program to be long-term does not mean it must go on indefinitely nor that it must never change. Periodic reviews should be conducted, and changes made to adapt rules and procedures as necessary. There can be a balance between certainty and flexibility if the program is assured to exist for some reasonable time (e.g., ten years) and if major changes in requirements are announced in advance with adequate time to phase in new approaches (e.g., five years). The appropriate time scales will of course vary with the nature of the change, with minor changes requiring little notice and major changes requiring enough time for decisionmakers and communities to plan and adjust.

Another long-term issue is environmental stewardship. The CDQ program as currently structured is, in large part, about economic development, but economic sustainability is dependent upon long-term assurance of a sound resource base—the fisheries. Thus, to be successful over the long-term the CDQ program will need to give more emphasis to environmental considerations.

While this report reviews the CDQ program in a broad way, there remains a need for periodic, detailed review of the program over the long term (perhaps every five years), most likely conducted by the State of Alaska. Such a review should look in detail at what each group has accomplished—the nature and extent of the benefits and how all funds were used. For a program like this, care must be taken not to use strictly financial evaluations of success. Annual profits gained from harvest and numbers of local people trained are valuable measures, but they must be seen within the full context of the program. It is a program that addresses far less tangible elements of "sustainability," including a sense of place and optimism for the future.

LESSONS FOR OTHER REGIONS

What emerges from a review of the western Alaska CDQ program is an appreciation that this program is an example of a broad concept adapted to very particular circumstances. In Alaska, where there were clearly definable communities, the fishery was already managed by quota with a portion of the quota held in reserve, and the communities had previous experience working within corporate-like structures. Others interested in the application of CDQ-style programs are likely to have different aspirations and different contexts. Wholesale importation of the Alaska CDQ program to other locales is likely to be unsuccessful unless the local context and goals are similar.

One region where the expansion of the CDQ concept has been considered is in the western Pacific, but such an expansion would need to be approached cautiously because the setting and communities are very different. The major differences between the fisheries and communities of the two regions are: the general lack of management by quota or total allowable catch (TAC) in the western Pacific; the pelagic nature of the valuable fisheries in the region; and the lack of clear, geographically definable "native" communities in most parts of the region. Application of the CDQ program to the western Pacific would require the Western Pacific Regional Fishery Management Council to define realistic goals that fit within council purposes and plans. Definitions of eligible communities would need to be crafted carefully so the potential benefits accrue in an equitable fashion to native fishermen.

Any new program, especially one with the complex goal of community development, should be expected to have a start-up period marked by some problems. During this early phase, special attention should be given to working out clear goals, defining eligible participants and intended benefits, setting appropriate duration, and establishing rules for participation. There should be real efforts to communicate the nature and scope of the program to the residents of any participating communities, and to bring state and national managers to the villages to facilitate a two-way flow of information. In addition to these operational concerns, those involved—the residents and their representatives—must develop a long-term vision and coherent sense of purpose to guide their activities.

1

Introduction

In 1992, the North Pacific Fishery Management Council (NPFMC) established a new program to help bring social and economic development opportunities to coastal villages in rural western Alaska. The new program, called the Community Development Quota (CDQ) program, allocates a portion of the annual fish harvest of certain commercial species directly to coalitions of villages, which, because of geographic isolation and limited access to sources of income, have had limited economic opportunities. In its first year, the council allocated 7.5 percent of the total allowable catch of Bering Sea pollock to be harvested exclusively by coastal communities of the Bering Sea region (or by those fishing partners authorized by the communities in exchange for royalties), and in subsequent years portions of other Bering Sea fisheries such as halibut, sablefish, crab, and assorted groundfish also were allocated to the communities. The goal is to enhance both economic and social development—that is, to help the communities develop the infrastructure and trained personnel necessary to support long-term participation in the fisheries industry and, along the way, build a stronger economic base and more vigorous communities.

In the Magnuson-Stevens Fishery Conservation and Management Act of 1996, Congress mandated that the National Academy of Sciences review the CDQ program in Alaska and evaluate its applicability to the western Pacific. A committee was set up to accomplish this task, and this report is the result of that review. Chapter 1 provides a brief description of the program, the committee's charge, and the approach taken. Chapter 2 describes the natural and social history of the Bering Sea region, while Chapter 3 provides a detailed description of the CDQ program. Chapter 4 provides the committee's review of the CDQ program, and Chapter 5 considers the advantages and disadvantages of a CDQ approach to fishery allocation and management. Chapter 6 addresses the applicability of the program to the western Pacific. Chapter 7 summarizes the study and offers recommendations.

THE BERING SEA FISHERY

The Bering Sea occupies an area of approximately 800,000 square kilometers, and is bounded by the Aleutian Islands on the south, the Bering Strait on the north, Russia to the west, and the coast of western Alaska (Figure 1.1). Largely due to the presence of an extensive continental shelf area, this region is one of the most highly productive marine systems in the world and supports some of the largest and most lucrative commercial fisheries conducted in U.S. waters. Commercial fishing operations harvest pollock, Pacific cod, sablefish, king and tanner crab, several flatfish species, rockfish, Atka mackerel, halibut, and salmon. The value of fishery harvest in the Bering Sea region was worth over $1.2 billion in 1996 (Alaska Fisheries Science Center, 1998).

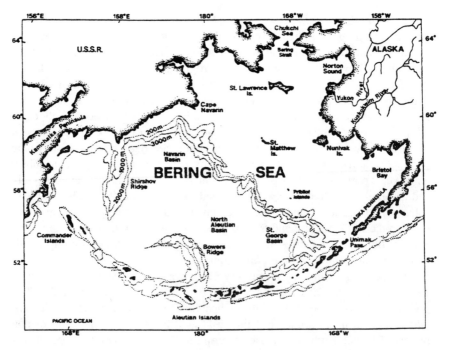

FIGURE 1.1   The Bering Sea occupies roughly 800,000 square kilometers, from the Aleutian Islands on the south, the Bering Strait on the north, Russia to the west, and the coast of western Alaska.

Until 1976, when the first steps toward creation of the exclusive economic zone (EEZ) of 200 miles were taken, not all harvests in the Bering Sea were managed, although several treaties attempted to bring some measure of control over harvest levels. Large foreign fleets dominated many Alaskan fisheries. But establishment of the EEZ brought a large portion of the Bering Sea under the jurisdiction of the United States. Congress delegated authority over the newly created zone to the NPFMC while retaining State of Alaska jurisdiction inside of 3 miles. The National Marine Fisheries Service (NMFS) manages the fisheries based on policy recommended by the council, including technical and enforcement activities.

Congress charged the NPFMC and other regional fishery management councils to recommend policy for governing the harvest, with the intent of regulating harvesting to sustainable rates. Once the Fishery Conservation and Management Act was passed in 1976, and the 200-mile zone was established, a transition from foreign fleets to American fleets began in the North Pacific Ocean fishery. Financial incentives such as low interest loans for building fishing vessels and processing facilities increased the pace of this transition. As the fisheries became more heavily capitalized, a dispute arose concerning the allocation of the pollock harvest between the fleet that used offshore processing and the fleet that used onshore processing facilities. NPFMC members from Alaska introduced the idea that part of the allocation of the pollock resource be awarded to communities of western Alaska, and this suggestion in time gave rise to the CDQ program.

Historically, the exploitation of Bering Sea resources from distant centers of industrial production has generated substantial wealth, but that wealth generally did not flow to indigenous people. While some Alaska Natives were participants in the development of this fishing industry, this is not true for many of the residents in the communities served by the CDQ program. The continued commercialization of Bering Sea marine resources tended to exacerbate the exclusion and marginalization of Alaska Natives from the commercial fishery and, to some degree, from their traditional resource-based lifestyle because, in part, they generally lacked the capital to participate. The concerns provided part of the impetus behind the creation of the CDQ program.

THE COMMUNITY DEVELOPMENT QUOTA PROGRAM

The Community Development Quota program is a unique quota-based fishing[1] management regime and not one of the traditional options used in fisheries management in the United States. In the CDQ program, a portion of the overall

---

[1]The committee uses the term "quota-based fishing" to describe harvest privileges that allocate a specific quota to an individual or community. The phrase "rights-based fishing" is more common in fisheries management, but the committee believes that term can be misleading. The concept of rights (versus privileges) can invoke an inappropriate sense of ownership or entitlement to a public resource.

total allowable catch (TAC)—or community development quota—is allocated to communities. The quota is not a total amount of fish but rather a share of the total annual amount allowed to be harvested—7.5 percent in the case of pollock at the time this study was undertaken.[2] The 7.5 percent CDQ pollock allocation is drawn from the "total allowable catch" (TAC) established annually by the North Pacific Fishery Management Council.

The nature of the CDQ allocation has both biological and political significance. Biologically, the CDQ allocation is not an allocation of additional harvest beyond the TAC. Politically, the CDQ allocation represents a reduction in the portion of the TAC that is available to the non-CDQ commercial fishery. Operationally, the CDQ allocation is drawn from a 15 percent harvest "reserve" established by the National Marine Fisheries Service as an aid to in-season management of the commercial fishery. Prior to the inception of the CDQ program, the full 15 percent reserve was, ultimately, released for harvest by industry. The CDQ allocation effectively reduced the reserve to 7.5 percent, thus ultimately reducing open-access harvests. In recent years, percentages of the halibut, sablefish, crab, and other groundfish fisheries have been allocated to the CDQ communities as well.

The CDQ quota is divided among six community organizations, which are themselves alliances of villages near the Bering Sea (Figure 1.2). Five of these six community organizations are nonprofit corporations organized under Alaska law. The sixth organization began as a for-profit corporation, but has changed its status to a nonprofit corporation.

Each CDQ group uses the royalty payments received for access to its share of the CDQ pollock quota as a source of funds for community development projects. The original policy of the NPFMC and subsequent policy require that the income gained be used on community development projects that tie directly to fisheries for fishery-related activities or to support education. Funds can, however, be invested outside the communities in financial instruments until they are ready to be used for community development. Projects funded by the CDQ quota can include construction and maintenance of infrastructure, such as ports and processing plants; purchase of fishing gear, such as lines, nets, and communications equipment, or investments in vessels; and training in fishing industry jobs, such as fish quality control. Both the terms of the sale of the CDQ quota and the community development plan that allocates the funds are overseen by the State of Alaska, with review by the National Marine Fisheries Service.

Details about the organization of the CDQ program are provided in Chapter 3. Some features of the CDQ program were based, in part, on lessons learned from the creation of village and regional corporations under the Alaska Native Claims Settlement Act (ANCSA) (43 U.S.C.A. numbers 1601-1628). In contrast

---

[2]According to Conference Report HR4328, effective 1/1/99 the pollock community development quota will be 10 percent of the total allowable catch.

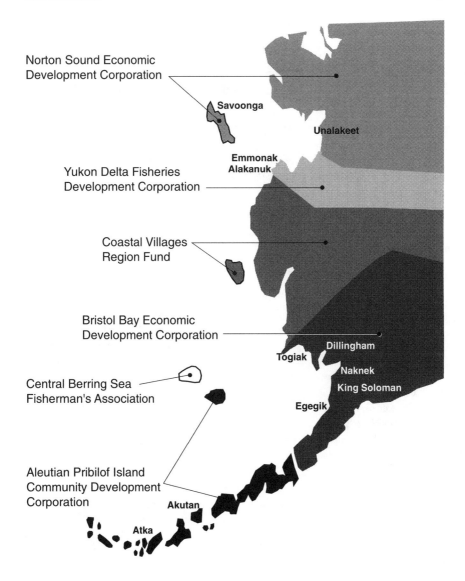

Norton Sound Economic
Development Corporation

Savoonga

Unalakeet

Emmonak
Alakanuk

Yukon Delta Fisheries
Development Corporation

Coastal Villages
Region Fund

Bristol Bay Economic
Development Corporation

Dillingham

Togiak

Central Berring Sea
Fisherman's Association

Naknek

King Soloman

Egegik

Aleutian Pribilof Island
Community Development
Corporation

Akutan

Atka

FIGURE 1.2 CDQ communities visited by committee and/or mentioned in the report.

to the ANCSA corporations, which have shares held and voted by individuals, the CDQ corporations are combinations of villages with the corporation boards of directors composed of representatives from the villages. A second critical distinction is the provision of oversight by the State of Alaska. The state approves community development plans and periodically reconsiders the allocation of the CDQ share of each fishery among the six CDQ groups, based on past performance and future plans.

THE COMMUNITY DEVELOPMENT QUOTA PROGRAM IN THE CONTEXT OF FISHERIES MANAGEMENT

The CDQ program is a management system that can be described as a limited access system. Limited access means that some people have recognized harvest rights while others do not (e.g., license limitations, attachment of harvest rights to tidal land ownership, and individual quotas). In the CDQ program, the allocation is made specifically to the communities and the share (although not the actual quota) is fixed unless regulatory policies change. The remainder of the pollock fishery in the Bering Sea is managed as a controlled open-access regime, in this case because the total catch is controlled by establishing seasons. During the open season anyone can fish and once the total allowable catch is caught, the season is closed. As a result of the time constraint, there is pressure on all fishers to participate aggressively in the limited open period, which leads to a "race for fish" during the open period and, over time as everyone competes to increase their take during the season, to the construction of more vessels to rapidly harvest the available fish. One effect of there being two different harvesting regimes is that the CDQ groups, which fish in a different time period, can get good prices for their share of the pollock. Efficiency may also be improved because the CDQ groups often form partnerships with vessels that have just completed fishing another season. These vessels take advantage of the previous season to get gear functioning and process lines working at best efficiency.

The use of the word "community" in the program title, "Community Development Quota," should not be taken to imply that the CDQ program is a type of community management. The CDQ program differs from community management, where communities have a direct role in decision making about management of the resource, such as timing of fishing seasons. The CDQ communities manage their business development plans and their harvest methods, but do not have a role in allocating quota. Quota is assigned by the North Pacific Fishery Management Council and the National Marine Fisheries Service.

THE COMMITTEE'S TASK

In the most recent reauthorization of the Magnuson-Stevens Fishery Conservation and Management Act (PL 104-297), Congress mandated that the National

Academy of Sciences conduct a review of the CDQ program in Alaska and evaluate the potential application of the program in the western Pacific (see Appendix A). The Committee to Review Community Development Quotas was established to perform this task. Specifically, the committee was asked to report on the performance and effectiveness of the community development quota programs that have been implemented under the authority of the North Pacific Fishery Management Council. For those Alaskan fisheries with adequate CDQ experience, the committee was asked to evaluate the extent to which such programs have met their objectives, such as providing communities with the means to develop ongoing commercial fishing activities, creation of employment, attraction of capital, development of infrastructure, and general promotion of positive social and economic conditions. The committee also considered what lessons could be learned to improve the design of new or proposed CDQ programs in Alaska and the western Pacific for accomplishing the purposes of CDQs. In essence, the committee was asked to judge whether the CDQ program has been effective in achieving its goals of economic development and thus, whether it warrants continuation or expansion.

The committee's members were selected to bring expertise in economics, marine ecology, anthropology, the fishing industry, sociology, Native Alaskan society, and other issues of relevance to this study (see Appendix B for biographies of committee members). No one on the committee has any direct relationship to the CDQ groups or to management of the CDQ program and all members and staff abided by standard National Research Council procedures to ensure the quality and independence of this report.

The charge to the committee occurred in the midst of a policy debate about how to address the overcapitalization caused by open-access fishing for pollock. In addition to the CDQ program, this debate has sparked careful analysis of the concept of individual fishing quotas as a particular way to limit entry and "rationalize" the fishery by reducing overcapitalization. The same legislation that created this committee also created the Committee to Review Individual Fishing Quotas to provide Congress with advice about the IFQ alternatives and that committee is expected to produce its report in late 1998. This report considers the relationship between individual fishing quotas (IFQs) and CDQs in Chapter 5.

THE COMMITTEE'S APPROACH

The CDQ program is new, and there is little literature available to form a basis for review. Therefore, the committee made a special effort to talk to people in the affected regions to learn about the program first-hand. The full committee held four meetings: Girdwood, Alaska; Seattle, Washington; Honolulu, Hawaii; and Molokai, Hawaii. Meetings were advertised in local newspapers, fishing industry trade publications, and on the World Wide Web site of the National Research Council. The committee received information from many sources (see

Appendix C), including invited speakers as well as interested parties at "open microphone sessions" at each meeting. Invited speakers included individuals such as state and federal fishery managers, inshore and at-sea processors, fishermen's organizations, Native Alaskan organizations, CDQ groups, and experts on community management. Additional written material was considered throughout the committee's deliberations.

In addition, small groups of committee members traveled to each of the CDQ regions and visited representative communities in each region and spoke directly with participants (Table 1.1). In most cases the committee members met with community leaders as well as with informal groups of community members. Finally, the committee contracted with a consulting anthropologist to spend time in several communities. The consultant talked in more depth with a wide range of the local populations about their participation in the program and their knowledge of the CDQ program and its benefits.

THE BROAD POLICY CONTEXT

To evaluate the CDQ program, it is important to understand the logic of U.S. fisheries policy over the past quarter century, especially four key elements. First,

TABLE 1.1   CDQ Locations Visited by Subcommittees

Date	Locations Visited
August 8, 1997	Nome, Alaska
August 9, 1997	Emmonak/Nome, Alaska Alakanuk, Alaska
September 9, 1997	Dutch Harbor, Alaska Akutan, Alaska Hooper Bay/Mekoryk, Alaska
September 10, 1997	Dutch Harbor, Alaska Atka, Alaska
September 15, 1997	Dillingham, Alaska
September 16, 1997	Togiak, Alaska St. Paul, Alaska
September 17, 1997	Naknek/Egegik, Alaska
September 18, 1997	King Salmon, Alaska

there has been a commitment to "Americanize" marine fisheries so that the benefits of ocean-based natural resources accrue to the American people rather than to foreign fishing nations. Second, within the mandate to Americanize coastal fisheries, there has evolved a presumptive entitlement for those individuals who have a historic tradition of using marine resources. Third, the development of specific fisheries policies has been motivated by a recognition of the potential benefits that may arise from regional control of natural resource use. Finally, there has been the affirmation that fisheries management programs offer important benefits—both socially and economically—to those local communities with a tradition of fisheries use. The establishment of the exclusive economic zone (EEZ) was the necessary condition to Americanize coastal and offshore fisheries resources. Within that new resource management regime, regional fisheries management councils were created to implement policies and programs in explicit recognition of: (1) historical use; (2) local participation in management decisions; and (3) the social and economic benefits to accrue to local people and places. In summary, the CDQ program in Alaska can be understood as a part of the Americanization of coastal fisheries resources.

For thousands of years, people of the north have lived off the sea. This exploitation was not limited to coastal resources; people traveled far out to sea to hunt and fish with many different types of capture devices. This region is home to one of the oldest and most sophisticated continuous maritime cultures in the world. This was not merely subsistence exploitation; the northern people have a long history of trade in marine resources.

During the past century, exploitation of Bering Sea resources from distant centers of industrial production has generated significant wealth. But because indigenous peoples were not typically participants in the development of the commercial fisheries, many did not have access to these economic opportunities. Nor were many Alaska Natives incorporated as labor into these fisheries. Moreover, some actions designed to protect marine species, such as the International Whaling Commission mandated cessation of North Alaskan bowhead whaling, limit the access of Alaska Natives to traditional sources of subsistence and trade goods.

The CDQ program can be seen as consistent with the general emphasis of American fisheries policy. The concern for management of aquatic resources and the key role of governmental organizations and institutions in providing program oversight are central components of the CDQ program. The primary emphasis on regional management places the CDQ program in the mainstream of contemporary fisheries policy.

The focus on the "community" as an important element in the CDQ program raises the obvious question as to what is meant by that term. As indicated above, the establishment of the EEZ introduced a form of community management into American fisheries. Prior to the EEZ, fishery resources falling between 12 and 200 miles of the American coastline were exploited by a variety of foreign nations. Fisheries management within 12 miles was the province of state and federal regulators. Disputes over fish harvesting fell to the State Department as part of U.S. foreign policy and was rarely addressed effectively.

With the coming implementation of the EEZ in 1976, all foreign fishing was brought under the control of American political structures. The fisheries management councils—consisting of industry and governmental representatives—became the operational means for managing coastal fisheries. The creation of the EEZ represented a transfer of wealth—in the form of the value of fish biomass—away from foreign fishing nations and to Americans who had previously fished, or who could claim some legitimate interest in harvesting activities in the newly established zone. Investments associated with this newly Americanized asset (the fish) continue to come from a range of nations, but the management of the fishery resources is now clearly under American jurisdiction.

Once the EEZ was in place, it became necessary to create management structures and processes under which the TAC for certain species could be determined and allocated among the newly endowed American fishing interests. The fishery management councils perform this management function. Potential council members are appointed by the Secretary of Commerce from lists provided by the governors of the coastal states from a variety of different fishing interests. For example, the North Pacific Fishery Management Council has members from the states of Oregon, Washington, and Alaska, including various industry groups based in these three states.

With the creation of the CDQ program, there is a specific asset allocation—allocation of biomass of various species to a specific group—this time to the residents of coastal Alaskan villages on the Bering Sea. With the creation of the CDQ program, there was a specific biomass allocation to a specific group—the residents of coastal Alaska—in the form of corporations composed of clusters of villages. Within this new regime, and under the general parameters established by the North Pacific Fishery Management Council and the state of Alaska, the groups of fishers themselves are responsible for the development of specific business plans for their share of the fisheries. They can choose to exploit the fishery themselves, or they can join with other partners in those endeavors. These parameters allow the CDQ associations to undertake particular managerial actions to support collective goals. Oversight by state and federal agencies assures that the well-being of the fishery resources takes precedence over other objectives. That is, economic development, job creation, or investment programs can only be implemented if they are consistent with the overarching goal of protecting the health and resiliency of the fishery resource.

The CDQ program can be seen as one component of a nested hierarchy of management structures and processes that operates in the North Pacific. It offers economic opportunity to residents of rural coastal Alaska that would otherwise be absent, while operating according to the biological imperatives developed and overseen by the North Pacific Fishery Management Council.

2

Description of the Region and Fishery

To evaluate the potential effects of the community development program on the communities in western Alaska, it is essential to first understand the underlying biological, social, and economic conditions. In many respects, these conditions are unique to the region, which have implications for the transferability of the Community Development Quota (CDQ) concept. This chapter provides an overview of the biological conditions of the Bering Sea fisheries, the social history of the region, and the structure and historical development of the fishing industry in the region. This sets the stage for more detailed discussions of the CDQ program (Chapter 3), the committee's evaluation (Chapter 4), and discussions of applicability of the approach in other regions (Chapter 5).

BIOLOGY

The Bering Sea is bordered by the Seward and Chukchi Peninsulas in the north, by the Kamchatka Peninsula in the west, and the Aleutian Islands in the south and southwest. This sea covers 3 million km^2, and one of its most unusual features is the extremely wide continental shelf, which makes the region an extremely productive ecosystem. The high productivity of the Bering Sea ecosystem exists despite the seasonal ice cover and limited light during the winter. Primary productivity on the southeast Bering Sea shelf is spatially variable and is highly episodic. Spring blooms are associated with the ice edge and with thermal stratification. In most oceanic ecosystems the primary production is consumed in the water column, and in many cases the nutrients are recycled within the top part of the water column. One unusual feature of the shallow Bering Sea shelf ecosystem is that much of the annual production escapes the water column consumers to feed a benthic system unusual in its amount of secondary productivity. Once it becomes part of the benthic system, this material is often slow to recycle because it becomes incorporated into long-lived benthic animals. An important additional factor in the high productivity of the southeastern Bering Sea shelf is the transport of nutrients onto the shelf from deeper water, which augments those nutrients generated locally (NRC, 1996; Sambrotto et al., 1986). This dependence could cause problems in the future for any change in the circulation of the Bering Sea, for example, a climate change could alter this transport.

The high primary productivity of the Bering Sea supports large numbers of birds, mammals, and fishes. There are some 50 commercially important fish species and at least 50 species of marine mammals. The fisheries in the region are some of the most abundant and productive in the world, especially for groundfish, halibut, salmon, and crab. Groundfish species in the Bering Sea include the walleye pollock, Pacific cod, several flatfish and rockfish species, and sablefish. Walleye pollock is an important species both as a predator and as a food source for other fish in its juvenile stage. Adult pollock are a major commercial asset for the United States, and they are marketed as fillets for a variety of products and as a minced and processed fish product known as surimi. Pollock roe are harvested during the winter, and are particularly valuable as an export to Asia, where they are considered a delicacy.

The increase in human activity and natural climate variability in the Bering Sea region have resulted in massive, if sometimes poorly documented, changes in the ecosystem over the last 50 years (NRC, 1996). Changes in the physical environment acting in concert with human exploitation of predators (whales, fish) have caused a shift in the abundance and distribution of many top predators and have caused pollock to dominate the ecosystem (NRC, 1996). These same changes also have resulted in dramatic fluctuations in the crab populations, as well as declines in some key marine mammal populations (NRC, 1996).

The domestic fisheries in the region are generally fully developed and most fisheries managers and economists consider them overcapitalized—that is, there are more boats and harvesting capability than available fish (NRC, 1996). Maintaining sustainable fishery populations, reducing bycatch of non-target species, and the minimizing negative impacts of the fishery on the marine mammal and sea bird populations are the most important biological issues that need to be addressed. These issues were of central concern to Congress in the 1996 reauthorization of the Magnuson-Stevens Act (i.e., the Sustainable Fisheries Act).

From the perspective of the CDQ program, some researchers contend that the ecosystem is very heavily exploited, and it seems extremely unlikely that the Bering Sea can sustain current rates of exploitation while also allowing the recovery of endangered species, especially large populations of marine birds and mammals (NRC, 1996). If the long-term goal of management is to maintain top-level predators, some fishing may have to be reduced. This emphasizes the need for

The sea has long been an essential source of food and other resources to the peoples of the Bering Sea region. In this 1984 photograph, Ella Tulik is drying herring using traditional techniques. (Photo by James Barker and provided courtesy of the Alaska State Council on the Arts, Contemporary Art Bank.)

for consideration of adaptive approaches to management (NRC, 1996) and may involve compromises that impinge on fishery resources. This issue is addressed in Chapter 4.

CULTURAL ASPECTS OF BERING SEA FISHERIES

The goal of the CDQ program is to improve the social and economic conditions in rural coastal communities. Understanding the relationship that has developed between Alaska Natives and the use of marine resources is a key component in evaluating the potential impacts of the CDQ program on these communities. This section provides background on these historical relationships, and is followed by a consideration of the ways a CDQ program could enhance these relationships.

Importance of Marine Resources to Native Communities

The prehistory and history of the indigenous peoples of the coastal margin and islands of the eastern Bering Sea are intimately tied to the utilization of the marine resources. This section provides a short introduction to the indigenous peoples and cultures of the eastern Bering Sea region, including a synopsis of the archeological evidence for occupation and resource use in the region, the linguistic and ethnic diversity of the region at the time of European contact in the 18th century, and some crucial cultural ideologies related to traditional use of resources. The historical changes associated with the coming of the commercial fishing industry to the region in the 20th century are also discussed.

Prehistory

Present information suggests that the earliest occupation of this coastal region occurred shortly after the onset of the Holocene and after sea levels had risen significantly (approximately 10,000 BP) (Table 2.1). The oldest sites of habita-

TABLE 2.1 Sites of Earliest Human Occupation of the Eastern Coastal Bering Sea Region

Area	Earliest Date	Site	Source
Aleutian Islands	8700 B.P.	Anangula Island	Laughlin, 1980
Alaska Peninsula	5100 B.P.	Ugashik Knoll	Dumond, 1984
Western Bristol Bay	4800 B.P.	Security Cove	Ackerman, 1964
Yukon-Kuskokwim Delta	1200 B.P.	Manokinak	Shaw, 1983
Norton Sound	4150 B.P.	Cape Denbigh	Ackerman, 1984
St. Lawrence Island	2100 B.P.	Punuk Island	Ackerman, 1984

tion in the region are found in the Aleutian Islands in nearshore coastal areas, but evidence for occupation in the other parts of the eastern Bering Sea indicate a gradual and uneven process, both in terms of population and the people's cultural adaptations to the area.

Since archeological research in this region is both difficult (particularly in the marshy delta of the Yukon and Kuskokwim Rivers) and limited, these dates should be regarded with caution.

There are several significant characteristics of the Bering Sea ecosystem that vary as one moves from north to south. One variation is the occurrence of winter pack ice. This occurs from the Bering Strait south to approximately Cape Newenham, but can extend further south to the vicinity of Port Moller during exceptionally cold winters (NRC, 1996). The Aleutian Islands, however, always remain ice free, a fact that allows the harvest of intertidal resources along the archipelago and thus provides a significant resource for inhabitants of the region (NRC, 1996). A second characteristic is the variation in the nature of the food sources available along the southern edge of the Bering Sea: both salmon and caribou, major resources available to mainland groups, are absent from or rare west from Unimak Island and on St. Lawrence Island (Laughlin 1980, Jorgensen 1990). Finally, virtually all of the societies of the eastern Bering Sea coast were dependent on the annual migrations of species that were abundant for relatively short periods each year, especially walrus, bowhead whale, salmon, and water-fowl. Harvests of these species focused on gathering a surplus that could be used during the winter and distributed at ceremonial feasts through a wide network of local kinsmen (Langdon, 1987a).

Strategies of Adaptation

Two general trends are apparent in the human occupation of the north: sedentary occupations came earlier in the south and adaptation strategies became increasingly complex through time. There are several distinct strategies of adap-tation to marine resources apparent in the record of cultural development. The first strategy of human adaptation to the eastern Bering Sea coastal environment, apparent at the Anangula Island site in the Aleutian Islands on the southern bound-ary of the eastern Bering Sea about 8,700 years ago, is one of a mixed subsistence (Laughlin, 1980). The inhabitants made substantial use of the rich intertidal resources of the area, including shellfish, sea urchins, chitons, seaweeds, birds, fish, and sea mammals. The second strategy, appearing on the Alaska Peninsula around 5,100 years ago, along the Brooks River drainage, is a riverine pattern in which salmon resources were apparently combined with caribou harvests to pro-vide sustenance (Dumond, 1984). A third strategy, found in the Cape Denbigh region in Norton Sound and initially used approximately 4150 B.P. relied on winter use of seals by hunting them through the pack ice (Ackerman, 1984).

With this innovation, the human occupation of the high Arctic became possible and human expansion pushed eastward to Greenland.

A fourth strategy was used on St. Lawrence Island, in the period around 2100 B.P. A major innovative leap was made allowing group harvesting and sharing of large sea mammals such as walrus and bowhead whale—a strategy used by the northwest Alaska Eskimo populations (Dumond, 1984). Finally, a diversified mobile strategy developed a little over a thousand years ago in the Yukon-Kuskokwim delta, combining the harvesting of many resources, including migratory waterfowl, small freshwater fish, small sea mammals, and herring in certain locations. This strategy made possible the permanent occupation of this last, difficult region (Shaw, 1983).

Linguistic Groups and Cultural Patterns

In the 18th century when the Russians, and later other Europeans, first came to the eastern Bering Sea, five linguistic groups were distributed along the coast and in the islands (Langdon, 1987a). These groups shared (to a certain degree) cultural strategies of communal interdependence and spiritual beliefs about the interdependence of animals and humans.

Aleut speakers (or Unangan—the term of self-identification in their language) occupied the entire Aleutian Archipelago as well as the Shumagin Islands in the North Pacific Ocean and the Alaska Peninsula east to Port Moller (Laughlin, 1980). West of Unimak Island, the Aleuts were predominantly marine mammal hunters (sea lions and harbor seals) and fishermen (halibut and cod). From Unimak Island eastward, they made use of salmon and caribou; marine mammal hunting and saltwater fishing were of less importance. The Aleut people were devastated by the Russian occupation in the 18th and 19th centuries, but they have persisted on their traditional lands in spite of various setbacks (Laughlin, 1980).

Alutiiq speakers (a Yup'ik Eskimo language spoken primarily on Kodiak Island and in the Prince William Sound region) occupied the Bering Sea shore of the Alaska Peninsula from just above Port Moller to about the Naknek River (Clark, 1984). The Alutiiq were primarily riverine salmon fisherman who also hunted small marine mammals (seals) and caribou. Alutiiq speakers are presently found at Meshik (Port Heiden) and Pilot Point, and they have strong cultural ties with communities around Chignik Lake and Chignik Lagoon (Clark, 1984).

Central Yup'ik speakers (practicing several dialectical variants) occupied the coast of Bristol Bay eastward and northward through the Yukon-Kuskokwim Delta region into the eastern portion of Norton Sound. These people used a variety of subsistence strategies depending on the abundance and availability of resources. Along the major rivers (Kvichak, Nushagak, Kuskokwim, and Yukon) they harvested and dried salmon and supplemented their diets with caribou, migratory waterfowl, freshwater fish, berries, and other resources (Van Stone, 1984). Along the coast away from the salmon streams, small marine mammal hunting was combined with herring fishing, freshwater fishing, migratory water-fowl hunting, and berry gathering (Van Stone, 1984).

Inupiaq speakers (northern Eskimo language) occupied the northern and western shore of Norton Sound and the small islands in the vicinity of Bering Strait (King Island, Sledge Island, and Little Diomede Island) (Ray, 1975). Along the mainland, groups combined fishing and hunting of small sea mammals (supplemented by an occasional beluga whale) and caribou. On the islands, Inupiaq speakers were predominantly marine mammal hunters who took walrus, bowhead whale, and seals and did some saltwater fishing for cod (Ray, 1975).

Siberian Yup'ik speakers occupied St. Lawrence Island and were in close contact with their Siberian Yup'ik residents on the Siberian mainland at East Cape and in periodic conflict with the Chukchi reindeer herders of the Chukotsk peninsula (Jorgensen, 1990). The Sivukaqmiut of St. Lawrence Island were large marine mammal hunters who acquired the majority of their foodstuffs and materials from harvests of walrus and bowhead whale. Seals and cod were supplementary resources (Jorgensen, 1990).

Cultural ideologies among all groups placed great importance on the acquisition of skills to enable them to effectively provide resources necessary to sustain their families. There was a clear division of labor between men and women. The acquisition of hunting and fishing skills was stressed among males. For females, skills in processing and storing food, sewing skins for clothing and for the covers of the vessels used by the men to hunt and fish were stressed (Clark, 1984). In the case of the Aleut, the predominant hunting orientation was the *baidarka* (kayak) manned by a single hunter (Clark, 1984). Among the Alutiiq and Central Yup'ik, the orientation was toward two men working as partners hunting together and sharing together as a unit, usually under the guidance of elders and shamans who shared their knowledge of the animal behavior to assist the younger hunters in their endeavors (Clark, 1984). Among the Inupiaq and the Siberian Yup'ik, coordinated units of 6 to 8 men successfully pursued large marine mammals such as walrus and bowhead whale using 20- to 30-foot open skin boats (Clark, 1984). Strategies combining competition and status with cooperation helped ensure the successful capture, landing, and rapid use of these large marine mammals.

Sharing and generosity were valued and practiced in these societies. Leaders sought to demonstrate their worthiness by sharing the products of their efforts widely among the people. The underlying spiritual and cosmological system of the groups stressed the interdependence of people and the resources that maintained them. Both human and animal life forms were thought to be cycled (reincarnated) from this world to the spirit world on death, and then returned from the spirit world to this world by birth (Fitzhugh and Kaplan, 1982; Fienup-Riordan, 1983). Animals were conceptualized as sentient non-human beings capable of recognizing human beings and their behavior. Animals made a conscious decision to deliver themselves to certain human groups based on their previous treatment by these groups (Fienup-Riordan, 1994).

Perhaps the most dramatic representation of the mutual dependence of humans and animals is demonstrated by the *Nakaciuq* (Bladder Feast) of the Central Yup'ik (Fienup-Riordan, 1983; Morrow, 1984). The Central Yup'ik believe that the spirit/life force of the seal is found in the bladder of each animal. When seals are killed, the bladders are stored in the *qasgiq* (ceremonial house) until the time of the feast in the ceremonial season (*Cauyarnariuq*—Season of the Drum) in the winter (Morrow, 1984). Then under the leadership of the shaman, the bladders, which are symbolic of wombs, are returned to the ocean where it is thought they will return to their homes beneath the ocean to be reborn. If they have been satisfactorily treated with respect by the humans, that is, the humans have demonstrated their worthiness, then the animals will return and once again give themselves to the human beings (Morrow, 1984). The animals judged the worthiness of humans by the humans' willingness to share the fruits of the animals' gifts among each other. This ceremonial giving became a source of cultural conflict when Euroamerican missionaries entered the central Yup'ik region and sought to modify their cultural pattern from one of communal interdependence to one modeled after Euroamerican and Judeo-Christian notions of familial independence. The European model never quite took hold, however.

The Modern Era

The various linguistic groups had remarkably different historical experiences since the coming first of European and later American influences to the region. Two sources of variation are particularly noticeable. The experiences of Alaska Natives are strongly affected both by the timing of initial Euroamerican contact and by the degree to which market-oriented economic development penetrated the community. The variations seen in these two dimensions are substantial and provide insight into the present conditions of peoples in the different areas.

Experience of the Unangan/Aleut

The Unangan were the first to encounter substantial European presence with the arrival of the Russian *promyshlenniki* (industrialists) in the middle of the 18th century. The Russians abducted Unangan wives and children, and thus coerced the Unangan men into harvesting sea otter and fur seal for the Russians. Through a combination of disease, warfare, and starvation, the Unangan were reduced from an estimated population of 12,000–15,000 to 2,000-3,000 in less than a century (Laughlin, 1980). Their traditional culture was radically altered and they took on a new identity, as Aleut, with the Russian Orthodox faith as a core foundation of the new culture. The culture evolved into a combination of marine mammal hunting for trade and subsistence purposes, and this strategy sustained Aleuts throughout their range until the latter part of the 19th century.

In the late 19th century, two commercial fishing enterprises based in the

Seattle, Washington, area began to appear along the western North Pacific Ocean coast and in the eastern portion of Bristol Bay. The first of these was the salt cod industry that began in the late 1870s, bringing Scandinavian dory fishermen to various fishing stations in the Shumagin Islands out to Sanak Island and up to Port Moller (Laughlin, 1980). Some of these fishermen remained and married eastern Aleut women, establishing new families in communities such as Unga, Pirate's Cove, Sanak, Pauloff Harbor, and Squaw Harbor (Laughlin, 1980). These fishermen and their industry did not penetrate as far west as Akutan, so that the western Aleuts retained the more traditional identity and cultural orientation.

The second commercial fishing development was the movement of the salmon processing industry, dominated by the canning sector, into Bristol Bay and the Alaska Peninsula in the late 1880s (Jones, 1976; and Van Stone, 1967). This was a massive incursion brought on by huge capital investments in plants, equipment, and vessels by Euroamerican firms. The canned salmon industry expanded as far west as False Pass, but the lack of significant salmon runs further out the chain limited the expansion in that direction. The canned salmon industry was well established along the east and north shores of Bristol Bay by the mid-1890s (Van Stone, 1967).

In 1942, the Japanese invaded the islands of the Western Aleutians and Pribilof Aleuts. Some residents from the westernmost inhabited Aleutian Island (Atka) were captured and transported as prisoners of war to Japan. The United States relocated other western Aleuts to camps in abandoned canneries in southeast Alaska, where they received poor housing and food. Meanwhile their communities were looted by U.S. troops, so that upon their return, their homes were virtually unlivable. This massive trauma was the subject of a major congressional hearing and reparations were subsequently paid to the western and Pribilof Aleuts for their mistreatment (Kohlhoff, 1995).

The impact of these cultural developments on later generations of Aleuts in the two areas encouraged the development of fishery-oriented communities, with males adopting commercial fishing as a new identity and lifestyle (Jones, 1976). In the post-World War II economy, the Aleuts in the east were able to position themselves as important participants in the salmon harvesting sector after the ban of floating fish traps in 1959. In the 1960s, the experience the Aleuts had gained in the waters of the central Gulf of Alaska provided them with a foothold when the crab industry began to expand. This commercial fishing heritage did not develop in the Aleut communities of Unalaska and west, nor in the Pribilof Islands.

Due to limited education and off-island exposure, few Pribilof Aleuts left the island permanently prior to World War II. Until the early 1980s, the Pribilof Island Aleuts continued their lifestyle tied to the fur seal harvest and processing, first for private companies and later as wards of the federal government (Jones, 1980). There was no development of commercial fishing or deep sea familiarity among the Pribilof Aleuts, although they did maintain some nearshore fishing for subsistence.

Experience of the Bristol Bay Yup'ik and Alutiiq

The effects of early Russian fur trading and later Euroamerican commercial fishing development on the Bristol Bay Yup'ik and Alutiiq provides a contrast to the Unangan/Aleut experience (Van Stone, 1967). On the eastern shore, early penetration of Alutiiq and several Yup'ik speaking villages by Russian fur traders led to substantial population reduction and a remnant population that adopted Russian Orthodoxy and in some cases took on a new identity as Aleut. This self-identification can still be seen in the 1980 and 1990 census records from the communities from Pilot Point to Iliamna. However, west of Lake Iliamna, where Yup'ik speaking and identifying groups occupy areas near the Nushagak, Igushik, and Togiak Rivers, more of their cultural heritage was preserved. The Nushagak River villagers are dialectically somewhat different from those to the west and term themselves the *Kiatagmiut*. From the village of Alegnagik westward, villagers speak Yup'ik forms more closely related to those of the Lower Kuskokwim groups. It appears that a wave of Lower Kuskokwim groups moved into western Bristol Bay following substantial population decline in that area in the early 19th century.

The canned salmon industry entered Bristol Bay in the 1890s to exploit the sockeye salmon runs returning to the lakes of the region. On the eastern shore, where development first took place, cannery owners were reluctant to hire Alaska Native workers because they could not leave if the work was not suitable; other workers were obliged to the canneries for transportation (Van Stone, 1967). The canneries in Dillingham and much later at Togiak, primarily used imported Italian and Scandinavian fishermen and Asian labor until World War II. It was only during World War II that Bristol Bay Yup'ik began to be used as fishermen and processing workers. Strict union policies and rampant racism led to conflicts in the post-World War II era as Italian and Scandinavian fishermen and Asian cannery workers sought to regain their preeminence. This led to the formation of two fishermen's organizations: one headquartered in Dillingham to represent the largely Yup'ik fishermen of western Bristol Bay and a second headquartered in Naknek to represent non-native fishermen who fished primarily in eastern Bristol Bay (Petterson et al., 1984).

The commercial fishing adaptation fit well with the continuing subsistence practices of the Yup'ik villagers. The Athapaskans from Lake Iliamna and Lake Clark began to enter the commercial fishery and followed a similar pattern. Men fished cannery-owned drift gillnet boats as teams, often consisting of brothers, father, and son or brother, and brother-in-law. Equal sharing of work, capital, and return was the Yup'ik formula that was readily adapted from the traditional male subsistence practices. Women worked in the cannery and later in the summer worked their own set net sites, selling part of the catch and drying and smoking the remainder. In August, they would leave the canneries and return up river to their villages where the last runs of dog and silver salmon would be harvested for subsistence.

The Bristol Bay pattern produced a sizable cash return and allowed purchase of fuel and food. In poor harvest years, canneries would extend credit for "grub stakes" to see their fishermen and families through the winter. In the winter, trapping and hunting, accompanied by seasonal feasting, particularly around the Russian Orthodox New Year, were the primary activities. Thus, in this instance there was an effective integration of traditional work patterns with commercial fishing activities, and even a transfer of skills between the two patterns (Langdon, 1987b). Despite problems associated with the sale of limited-entry fishing permits, the integration of commercial fishing activities with traditional subsistence practices continued to be a major foundation of Yup'ik society and values in western Bristol Bay into the 1980s (Langdon, 1991).

It is important to note that the crab fisheries in the 1960s and 1970s were conducted by vessels over 60 feet in length based primarily out of Seattle and Kodiak. Unlike the Alaska Peninsula Aleut, the Bristol Bay Yup'ik did not have large vessels or enough deep water experience to compete on relatively equal footing in these new industries.

With the creation of the EEZ and the exclusion of foreign fishing, a sac roe herring fishery developed in the Togiak district of western Bristol Bay. Euroamerican purse seiners from districts in other parts of Alaska sought to monopolize this new industry to the exclusion of the local Yup'ik gillnet fishermen. However, the Yup'ik responded and won an important court decision that allowed foreign vessels to purchase sac roe herring from them, since domestic processors refused to purchase Yup'ik production (Langdon, 1982).

The restriction of entry into the Bristol Bay salmon and sac roe herring fisheries is a clear case of the denial of opportunity to local fishermen through arrangements between processors and groups of Euroamerican fishermen (Langdon, 1982). In many cases, local fishermen did not participate in the fishery unless outside labor was not available, and little effort was made to include them in the industry (Van Stone, 1967).

Experience of the Yukon and Kuskokwin Yup'ik

North of Bristol Bay, near the lower reaches of the Yukon and Kuskokwim Rivers, the experience of Alaska Natives was again different from that in other regions. In this region, significant and sustained contact with Euroamericans did not occur until well into the 20th century. Waters of the delta are extremely shallow, which was a major impediment to sailing vessels used by early fur traders (Oswalt, 1990). In the latter part of the 19th century, expansion of the salmon canning industry stopped near Bristol Bay due to the high cost of production on the Kuskokwim and Yukon Rivers. These factors combined to keep the regions along these rivers free from significant external influences until the mid-20th century when improved transportation (airplanes) and expanded markets allowed a skiff-based commercial fishery to be developed. In the Yukon and Kuskokwin delta, an opportune fusion of skills, seasonal patterns, and working style similar to that in Bristol Bay allowed Yup'ik populations to develop a mixed subsistence-based society that provided an improved standard of living while retaining strong subsistence orientations, and values (Wolfe, 1984; Fienup-Riordan, 1986b, 1994). In some of the villages in the Yukon-Kuskokwim delta, local fishermen received permits and were able to participate in the fisheries. In other instances, some local fishermen were excluded from developing commercial fisheries because those fisheries were considered to be part of another river system. However, the limited size of commercial resources and distance from markets meant a lower income for Yukon and Kuskokwim fishermen compared to Bristol Bay fishermen (Langdon, 1991). The Yup'ik people of the Yukon-Kuskokwim delta have for over 25 years acted to protect their subsistence values and resources for future generations.

Experience of the Norton Sound Yup'ik

The experience of Yup'ik cultural change in Norton Sound is similar to that of the Yup'ik to the south; however, the gold rush on the Seward Peninsula that led to the creation of Nome had significant effects on cultural development. Nome has been the center of significant disruptive influence on local societies since its founding. Alcohol, violence, and ethnic conflict have been central factors in the social disruption. It is important to note that Alaska Native institutions (regional corporations, school districts) have elected to locate their headquarters in Unalakleet rather than in Nome. Many Alaska Natives feel that the Euroamerican financial power structure that has substantial influence in Nome is not supportive of their goals.

A limited commercial fishery for salmon, sac roe herring, and king crab appeared in the Norton Sound area in the 1960s. These fisheries were conducted using gillnets and small boats similar to those in the Yukon-Kuskokwim region. The more northerly location of these fisheries placed them at the margin of the economy's ability to absorb their production and provide the fishermen with a reasonable rate of return. Commercial fisheries have been operated as far north as Kozebue, but they have not been profitable over the long term due to poor weather and low prices. Low run size and extreme population fluctuations also have combined to limit commercial fishing as a viable source of income in certain years. Subsistence activities continue to be important to the Inupiaq and Yup'ik residents of the region. As in regions further south, the mixed subsistence-based economy includes wage labor and transfer payments as important components of the local economy.

Experience of St. Lawrence Yup'ik

The Siberian Yup'ik people of St. Lawrence Island have a history distinct from other groups in the region. The population of the island was reduced by starvation from approximately 1,500 to 300 in 1878-89 (Bockstoce, 1986). A primary cause of this starvation was the decimation of the walrus and bowhead whale herds, the two primary staples of the *Sivukaqmiut* (Siberian Yup'ik term for residents of the island), by Yankee whalers in the 1860s and 1870s. The remaining Yup'ik population consolidated at the oldest settlement on the island, *Sivukaq* (now called Gambell). In 1892, reindeer were introduced at the Teller Reindeer Station under the sponsorship of Sheldon Jackson, a missionary working to improve local economic conditions (Jackson, 1906). The native community on St. Lawrence Island has retained its essential subsistence character and is considered a center of cooperative group marine mammal-hunting activities. While there has been some subsistence harvesting of cod traditionally and into the contemporary period, commercial fishing had not been practiced by residents of St. Lawrence Island until it was initiated under the CDQ program in 1996.

INDUSTRY STRUCTURE AND HISTORY

To evaluate the CDQ program and its effects on the participating communities, some understanding of the history and structure of the commercial fishing industry of the Bering Sea and Aleutian Islands (BSAI) is necessary. Historically, the management system adopted by the North Pacific Fishery Management Council (NPFMC) has separated halibut, sablefish, crab, and groundfish fisheries, and these fisheries have shown differences in management structure, as well as historical development of the fleets. The following sections describe some of the key elements of these fisheries. These sections provide a general introduction into both the complexities of fishery management in the North Pacific and the historical development of the domestic commercial fisheries in the Bering Sea. This introduction provides context for understanding the CDQ program. Several aspects of the historical development of the industry are particularly relevant, including who participated, the relative recency of the domestic industry, how the domestic industry was actively "developed," and the extreme variability of some of the fishery resources involved.

The Commercial Halibut Fishery

The commercial halibut fishery in the Bering Sea and Aleutian Islands targets a single species, the Pacific halibut (*Hippoglossus stenolepis*) and, by regulation, is conducted exclusively with hook-and-line gear known as longline or setline gear (Figure 2.1). In 1996, 313 vessels harvested 5.27 million pounds of halibut from Area 4, the overall management unit established by the International

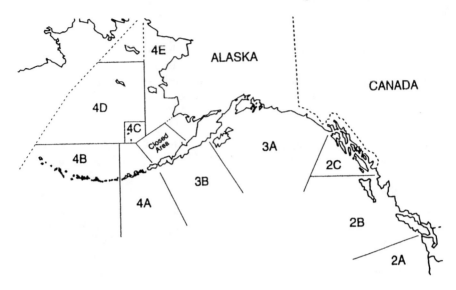

FIGURE 2.1   International Pacific Halibut Regulatory Areas for the Alaskan halibut fishery.

Pacific Halibut Commission for the BSAI areas (International Pacific Halibut Commission (IPHC), 1997). These figures include harvests from both the individual fishing quota (IFQ) program and the CDQ halibut fishery. The IFQ and CDQ programs were introduced into the halibut fishery in 1995. Prior to the 1995 season, the halibut fishery was conducted as an open-access fishery. Increasingly short seasons, overcapitalization, concerns over safety, and other factors had been concerns in the halibut fishery since shortly after the implementation of the Magnuson-Stevens Act in 1976 (NPFMC, 1998). Efforts at limiting access in the fishery led to a number of proposals, and eventually to the implementation of the Alaskan halibut (and sablefish) IFQ program in 1995 (NPFMC, 1998). The Alaskan halibut IFQ program is designed to allocate a percentage (quota share) of the total allowable catch to individual fishermen who qualify to receive or purchase quota share based on criteria established by the North Pacific Council. The history of the fishery in the BSAI was previously summarized by the NRC (1996):

> Halibut were reported in the Bering Sea by U.S. cod vessels as early as the 1800s. However, Bering Sea halibut did not reach North American markets until 1928 (Thompson and Freeman, 1930). Small and infrequent landings of halibut from the Bering Sea were made by U.S. and Canadian vessels between 1928 and 1950, but catches were not landed every year until 1952 (Dunlop et al., 1964). The catch by North American setline vessels increased sharply between

1958 and 1962 (exceeding 3,300 t), then fell to a low of 130 t by 1973 before recovering to a high in 1987, and then declined slowly.

While there are many participants in the halibut fishery, most of the catch is captured by a smaller segment of the fleet. The halibut fishery can be prosecuted by small vessels, but the catch is concentrated in the larger vessels in the fleet. Although a majority of vessels participating in the fishery were 35 feet in length or less, 76 percent of the 1996 combined IFQ and CDQ Area 4 halibut harvest was captured by the 33 percent of the fleet consisting of vessels greater than 55 feet long (IPHC, 1997). Similarly, although vessels from many CDQ villages adjacent to Area 4 participated in the open access fisheries (particularly Atka, Akutan, Unalaska, St. Paul, St. George, Mekoryuk, Toksook Bay, and Tununak), the initial distribution of quota shares in the IFQ program reflected catch history during the qualifying years, and local residents did not receive a significant share of quota (CFEC, 1996). Exceptions to this pattern were found in sub-areas 4C and 4E.

Notably, these are sub-areas that received specific local fishery development attention from the International Pacific Halibut Commission and the North Pacific Fishery Management Council prior to the inception of the IFQ program. In the late 1980s, the commission and the council employed a combination of trip limits, short openings, and vessel clearance requirements to "provide more fishing opportunity for local fishermen" from the Pribilof Islands (Area 4C) and Nelson and Nunivak Islands (Area 4E) relative to non-local fleets (Hoag et al., 1993; see also 53 FR 8938, March 18, 1988, 53 FR 10536, April 1, 1988, and 53 FR 20327, June 3, 1988).

Some residents of the CDQ communities have participated in the halibut, sablefish, crab, and groundfish fisheries as crew members, skippers, or vessel owners. However, there had not been any formal participation in the Bering Sea fisheries by these communities as a whole prior to the implementation of the CDQ program. With the implementation of the halibut and sablefish IFQ program, data were collected indicating the residency of the individuals involved in those fisheries. Similar information is more difficult to obtain for the crab and groundfish fisheries. In general, residents of the CDQ communities have had a longer and more extensive participation in the halibut fishery than the other Bering Sea fisheries.

Sablefish

The domestic commercial fishery for sablefish (*Anoplopoma fimbria*) in the Bering Sea and Aleutian Islands is a relatively recent development. Domestic landings were first recorded in 1980 and foreign harvests were not eliminated until 1988 (NPFMC, 1989a). This pattern of recent domestic fishery development contrasts sharply with both the halibut fisheries of the region and the domestic sablefish fishery in the Gulf of Alaska. Domestic sablefish fishing in the Gulf began in the early 1900s, was displaced by foreign harvests in the early 1970s, and resumed prominence in the mid 1980s as foreign harvests were phased out (Low et al., 1976; NPFMC, 1989a). Japan dominated the foreign harvests of sablefish throughout the North Pacific.

The domestic sablefish fishery is prosecuted with longline gear similar to that used in the halibut fishery but because sablefish are generally found at much greater depths and farther offshore, the sablefish fishery is prosecuted using relatively larger vessels. Participation by local residents of the Bering Sea and Aleutian Islands areas was very low prior to the inception of the CDQ program and contrasted sharply with local participation in the halibut fishery (NPFMC, 1992).

Commercial Crab Fisheries

The commercial crab fisheries covered under the council's BSAI king and Tanner crab fishery management plan target king, Tanner, and snow crabs. Targeted king crab are of the family Lithodidae: red king crab (*Paralithodes camtschatica*), blue king crab (*P. platypus*), and brown or golden king crab (*Lithodes aequispinus*). The Tanner crab fisheries involve Tanner or bairdi crab (*Chionoecetes bairdi*) and to a much lesser extent grooved Tanner crab (*Chionoecetes tanneri*) and triangle Tanner crab *(Chionoecetes angulatus)*. The snow crab fishery targets snow, or opilio, crab (*Chionoecetes opilio*).

The commercial crab fleet is characterized by vessel type: whether the vessel delivers live crab to a processing plant (a catcher vessel) or processes the crab at sea on board the vessel while harvesting (a catcher-processor). The fleet is also characterized by vessel length and by residency of the vessel owner. Currently, the BSAI crab fisheries are constrained by a moratorium on vessel entry and a more restrictive license limitation program is pending implementation. The council's license limitation program for the BSAI crab fisheries is expected to issue species- area-specific licenses to 427 distinct crab vessels. Of these 427 vessels, 400 are catcher vessels and 27 are catcher-processors. The majority of the catcher vessels are less than 125 feet in overall length, whereas all but 2 of the catcher-processor vessels are over 125 feet long. Only 175 of the 400 catcher vessels are reported to be owned by Alaskans, while residents of Washington own the bulk of the remaining vessels. The majority of the crab catcher-processor vessels is owned by Washington residents. With the exception of the recipients of licenses for Norton Sound red and blue king crab, virtually none of the projected recipients are from coastal communities adjacent to the respective BSAI crab fisheries (all data from NPFMC, 1997a).

The BSAI crab fisheries vary considerably by number of participating vessels, harvest amount, value, and season length as shown in Table 2.2.

Although there is some overlap between the crab and groundfish fleets, a substantial portion of the BSAI crab fleet is comprised of dedicated crab vessels.

TABLE 2.2 Participation Levels, Harvests, Ex-Vessel Values, and Season Lengths for Selected Crab Fisheries in the Bering Sea/Aleutian Islands in 1996

Fishery	Year	Vessels (Days)	Harvest (Value)	Ex-Vessel (Value)	Season Length (Days)
Bristol Bay red king	1996	196	8.4	33.6	4
Pribilof red & blue king	1996	63	1.1	3	11
St. Matthew blue king	1996	122	3.1	2.2	8
Adak brown king	1995/96	25	4.9	9.6	289
Bering Sea Tanner	1996	195	1.8	4.5	16[a]
Bering Sea snow	1996	234	65.7	85.6	45

NOTE: Harvests in millions of pounds of live crab. Ex-vessel value in millions of dollars.
[a]Bering Sea Tanner crab season length represents 4 days as bycatch during Bristol Bay red king crab season plus 12-day directed season following closure of the red king season.
SOURCE: ADF&G, 1997

Within the crab fleet prior to the license limitation, there was considerable overlap between the vessels that fished the Bristol Bay red king crab and the other crab fisheries, with the exception of the Adak brown king crab and the Norton Sound red and blue king crab fisheries. The Adak fishery is informally an exclusive fishery by virtue of physical conditions (depth and tidal currents) and season length. The Norton Sound fisheries are exclusive by regulation (vessels that register to fish in Norton Sound cannot fish in any other area).

The following historical account of commercial crab fishing in the BSAI draws directly on the Council, 1989 (Zahn, 1970; Otto, 1981; ADF&G, 1997; and Browning, 1980). Japanese tangle-net fishing for king crab in the eastern Bering Sea began in 1930, ended in 1940, and resumed in 1953. A Russian king crab fleet entered the eastern Bering Sea fisheries in 1959. Domestic U.S. exploration of a potential commercial fishery began (and ended) in 1941 and was resumed in 1946 utilizing trawl gear. Domestic effort increased at the end of the decade, but it declined in the late 1950s, and no domestic catch was recorded in 1959. Domestic king crab harvests in the Bering Sea were low and variable until the mid-1960s. Management of the domestic fishery initially was vested in the U.S. Bureau of Commercial Fisheries (the precursor to the National Marine Fisheries Service), but was transferred upon statehood to Alaska's Board of Fisheries and the Alaska Department of Fish and Game. The State of Alaska continues to manage the crab fishery under authority delegated to it by the North Pacific Council.

From the beginning, the Board of Fisheries moved to "protect local fleets" (NPFMC, 1989b). In 1964, the U.S. government negotiated bilateral agreements with Japan and the USSR with the aim of gradually supplanting the foreign fisheries with domestic harvesting and processing capacity. Russian and Japanese king crab harvests ended in the 1970s. At the same time, domestic production in the Bristol Bay red king crab fishery increased throughout the 1970s until in 1979, the port of Dutch Harbor/Unalaska became the number one U.S. port in terms of dollar value of commercial fishery landings. The steady progression of record red king crab catches peaked with the 129.9 million pounds harvested during the 40-day 1980 season. The red king crab fishery then experienced dramatic decline, and there was no commercial fishery in 1983. The fishery reopened in 1984, and catches rose to 20 million pounds in 1990, but declined again with a total closure of the Bristol Bay red king crab fishery in 1994 and 1995. The red king crab fishery was reopened in 1996.

Compared to the Bristol Bay red king crab fishery, all of the other crab fisheries in the BSAI are relatively recent. Despite later development, most of the other crab fisheries have shared the "boom and bust" pattern of the Bristol Bay red king crab fishery. A directed domestic Tanner crab fishery began in 1974. Foreign Tanner crab fishing was phased out by 1980. The record harvest of Tanner crab in the Bering Sea, 66.6 million pounds, occurred during the 1977-78 season. The Tanner crab fishery was closed in both 1986 and 1987 due to stock declines, and more recently in 1997. The domestic snow crab fishery began in 1977. The record harvest for the snow crab fishery, 328.6 million pounds, occurred in 1991, but catches have steadily declined since then to a catch of 65.7 million pounds in 1996. The commercial Pribilof blue king crab fishery began in 1973, was closed from 1988 to 1994, and reopened in 1995. The commercial Pribilof red king crab fishery first opened in 1993.

Commercial Groundfish Fisheries

The BSAI Groundfish Fishery Management Plan defines the following target species for which specific total allowable catch (TAC) levels are set annually: pollock (*Theragra chalcogramma*), sablefish (*Anoplopoma fimbria*), Pacific cod (*Gadus macrocephalus*), squid (*Berryteuthis magister* and *Onychoteuthis borealijaponicus*), yellowfin sole (*Limanda aspera*), rock sole (*lepedopsetta bilineata*), flathead sole (*Hippoglossoides ellassodon*), Atka mackerel (*Pleurogrammus monopterygius*), Greenland turbot (*Reinhardtius hippoglossoides*), Pacific ocean perch (*Sebastes alutus*, also includes the "other red rockfish" complex), arrowtooth flounder (*Atheresthes stomias*), other flatfish (primarily by Alaska plaice— *Pleuronectes quadrituberculatus*), and other rockfish (mostly by shortspine thornyheads—*Sebastolobus alascanus*).

Various fleets pursue the species of these fisheries, but, unlike the crab and halibut fisheries, some of the groundfish fisheries of the BSAI are prosecuted using multiple gear types (e.g., Pacific cod, Greenland turbot, and sablefish), while the others are prosecuted with trawl gear exclusively. Within the trawl fisheries, some (e.g., rock sole) are pursued almost exclusively by catcher-processor vessels. In contrast, the pollock fishery (an exclusively trawl fishery) is allocated between offshore processor vessels (catcher-processors or offshore motherships receiving deliveries from catcher vessels) and inshore processors (shore-based plants and nearshore floating processors receiving deliveries from catcher vessels). The Pacific cod fishery is allocated among trawl, longline, and pot gears, and the catch is processed by both inshore and offshore processors. Finally, some (e.g., arrowtooth flounder) "target" species are hardly pursued as targets at all.

Although multiple gear types are present, trawl gear dominates the ground-fish fisheries of the BSAI, accounting for 92 percent by weight of all groundfish harvested in 1995 (Kinoshita et al., 1997). The trawl fleet itself is divided into catcher vessels and catcher-processor vessels, with the latter category further divided by the type of product produced. In order of decreasing vessel size, the trawl catcher-processor fleet includes surimi processors (sometimes in excess of 300 feet) that also sometimes produce fillets; fillet processors (220-300 feet); and head and gut processors (20-220 feet). Four shore-based processors account for most of the "inshore" processing of BSAI groundfish; three plants are located in Unalaska and one in Akutan.

Pollock is the dominant groundfish fishery both by value and weight. Collectively, the 1995 groundfish fisheries landed 1.93 million metric tons worth an estimated $585 million in ex-vessel value (Kinoshita et al., 1997). Of these amounts, pollock represented 69 percent by volume and 48 percent by value. The groundfish fisheries are subject to a license limitation plan approved by the Secretary of Commerce; the plan excludes the longline portion of the sablefish fishery (which falls under the IFQ program) and a small vessel jig-gear-only fishery for Pacific cod in the Aleutian Islands area established by the Council shortly before adoption of the license program.

Analysis of the license program estimated that 407 catcher vessels and 141 catcher-processors (fixed and trawl gear) would receive licenses to fish groundfish in the BSAI, including vessels jointly licensed to fish in the Gulf of Alaska groundfish fisheries (NPFMC, 1997a). The analysis also indicated that most of the harvesting capacity in both the catcher vessel and the catcher-processor sector resides in the larger vessels in each sector, and that both vessel sectors are based predominantly in Washington State (NPFMC, 1997a). The analysis also underscored the fact that the BSAI groundfish fisheries are highly industrialized and overcapitalized. The state of the groundfish fisheries results directly from the fisheries development policies pursued in the BSAI.

Although domestic commercial groundfish fishing occurred in the BSAI as early as an 1864 venture that fished for Pacific cod (Cobb, 1927), the domestic groundfish fisheries were virtually non-existent when the Fishery Conservation and Management Act was passed in 1976 (Rigby, 1984). Domestic harvests in the BSAI in the late 1970s primarily consisted of a bait fishery supplying the then soaring king crab fishery (BSAI FMP, p. 8). When the Act was passed in 1976, groundfish fishing in the BSAI was a foreign activity (principally by the Japanese and Soviet fleets, but fleets from Poland, Korea, and Taiwan were also active). The Act was amended in its second year to emphasize that "a national program for the development of fisheries which are underutilized or not utilized by the United States fishing industry, including bottom fish off Alaska, is necessary. . ." (Sec. 2(a)(7)). Domestic harvesting capacity responded to the emphasis on displacing foreign harvesting fleets by utilizing foreign processing fleets during the rapid buildup of capacity that occurred during the so-called "joint-venture era."

Development of the groundfish fisheries was accelerated further by additional Congressional attention to the processing sector as reflected in subsequent amendments to the Act, including the processor preference amendment and the American Fisheries Promotion Act. The former set the stage for the beginning of the end of the joint ventures, while the latter formalized the informal policy of maintaining some foreign access to harvests in exchange for highly desired technology transfer and opening of export markets to assist in the effort to "Americanize" fisheries like the groundfish fisheries (specifically, the BSAI pollock fishery). In addition to this external focus on foreign nations, Congressional fishery development policy also had an internal focus: tax credit and loan guarantee programs were provided to encourage investment in the domestic industry. The rapid success of these development policies can been seen in the line graphs presented in Figure 2.2 below:

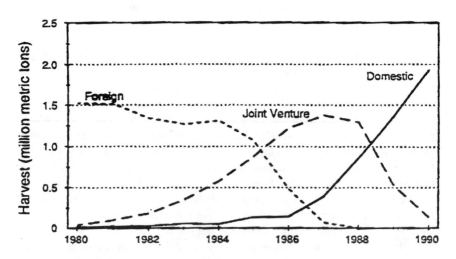

FIGURE 2.2   The "Americanization" of the Alaska groundfish harvest.

NOTE: The graphs pertain to all groundfish harvests off Alaska (i.e., the Gulf of Alaska plus BSAI harvests, but the scale of the BSAI groundfish fisheries is such that the fisheries are effectively drowned out in the graphs). SOURCE: NPFMC, 1990.

Like the crab and halibut fisheries, only more so, the groundfish fisheries of the BSAI are a direct product of government-directed fisheries development policies. In this, they provide a suitable context for a review of the CDQ program. The history of the groundfish fisheries is a history of development pursued through the allocation of economic opportunities. The EEZ, the Processor Preference Amendment, the American Fisheries Promotion Act, the creation and subsequent closure of joint ventures, the individual fishing quota (IFQ) program, the inshore/offshore allocation, the cod trawl/fixed-gear allocation, the small boat jig-gear allocation and the CDQ program are all acts by Congress or the Council in pursuit of particular visions of "development." [1] The fact that government-backed development is the consistent theme of all post-Magnuson Act fisheries management is often overlooked by participants in north Pacific fisheries. A quote from the not-so-distant past serves as a reminder that those persons embroiled in debate over today's development policies were often endowed by past development policies:

> It's almost like a battlefield, the U.S. exclusive economic zone (EEZ) of Alaska, with the North Pacific Fisheries [sic] Management Council acting as arbitrator among warring factions of seafood-hungry nations. The council ruled in our favor this year, eliminating all foreign fishing from the zone. . . And with less of their own production to rely on from the Bering Sea, both Japanese and Korean buyers will be clamoring for groundfish from the U.S. zone. (Seafood Business, 1988)

Current Issues in Western Alaska

The economic conditions in many Alaska Native communities reflect the lack of recent commercial development. Although the subsistence economy still works in some communities, the lack of economic opportunities for Alaska Native continues to exist. The lack of economic opportunities and the despondency

---

[1]The EEZ, the IFQ program and the inshore/offshore allocation are described elsewhere in this report. The Processor Preference Amendment (PL 95-354) was fashioned by Congress during the 1978 amendments to the MSFCMA and spelled the beginning of the end for the "joint venture" production in which harvests by a domestic fleet were processed on-board foreign at-sea processors by establishing a priority for domestic processors. The American Fisheries Promotion Act (PL 96-561, the 1980 reauthorization of the MSFCMA) was another step in the Congressional effort to "Americanize" fisheries in the EEZ and formalized the "fish and chips" policy of encouraging foreign investment in domestic processing capacity. In late 1990, the Council "brought to an end the joint venture era in the groundfish fisheries off Alaska by apportioning all quotas [the 1991 total allowable catches by fishery] to fully domestic fisheries" (NPFMC, 1990). In 1993, the Council divided the harvest of Pacific cod in the Bering Sea and Aleutian Islands management areas between contesting gear factions by establishing specific trawl gear and fixed gear (longline and pot gear) allocations. To support a fledgling small boat cod fishery developing in the Aleutian Islands area, the Council also established a third direct allocation of cod reserved for small boats exclusively using jigging gear.

that results has had profound effects on the social conditions of these communities and has exacerbated existing social problems (Alaska Natives Commission, 1994). As an example, alcoholism and suicide among Alaska Natives continues to remain high, and recent studies suggest that the rates for alcoholism, alcohol-related deaths and injuries, and suicide are significantly higher in western Alaska than in the general U.S. population (Gessner, 1997; Landen et al., 1997). Some villages have attempted to counteract high rates of alcoholism by restricting the sale of alcohol in the villages, and those policies have reduced the rates of alcoholism and alcohol-related deaths (Landen et al., 1997).

These social problems, in addition to poor sanitation and immunization rates, and the limited availability of health care, education, and employment opportunities have been well-documented (Alaska Natives Commission, 1994). In many cases, Alaska Natives have among the lowest levels of income, employment, and life expectancy of all ethnic groups in the United States yet have among one of the highest birth rates (Hensel, 1996; Alaska Natives Commission, 1994). These conditions contribute to a reduction of the ability of villagers to participate in many of the subsistence activities that provide important food and domestic products for these regions, and further exacerbate economic and social problems (Marshall, 1988).

Some villages have generated locally based businesses and industries that have been important sources of employment and hope through coordination among different organizations (e.g., Emmonak, Saxman in Southeast Alaska). Generally, there has not been an expansion of locally-based businesses and industries (Alaska Natives Commission, 1994). All of these factors indicate that the lack of development continues to be a central factor in the social and economic problems facing western Alaska communities.

What is Development in Western Alaska?

The idea of "development" has been evolving in the past several decades, both among some academic experts and among the peoples concerned. Of course, the alleviation of poverty and the attendant social ills remain fundamental and are the common denominator of all definitions. But the narrow view of economic growth—expansion of production, productivity, and per capita income—as the greatest good and an end in itself has been superseded by, or included in, a more general understanding of development as the enrichment of chosen forms of human life. It is material growth that now appears as the means. The end is the overall fulfillment of human existence according to a people's own cultural conceptions of what a good life is. Development currently includes the concepts of the enrichment of a way of life and self-determination, concepts beyond a strictly economic perspective. No doubt many factors have contributed to this shift in perspective. Among them is the recognition of cultural diversity. Economists and other social scientists long thought that development was impeded by the

The idea of "development" has been evolving, and is no longer focused exclusively on economic growth. The goal of development initiatives such as the CDQ program includes both alleviation of poverty and attendant social ills and enrichment of a way of life and self-determination. (Photo of Charles Hanson by James Barker and provided courtesy of the Alaska State Council on the Arts, Contemporary Art Bank.)

indigenous culture of the so-called traditional societies. Yet often what they were witnessing was the integration of modern technologies, goods and relations in and by local traditions, which gave such "development" (as locally defined) an unrecognizable appearance. In many situations cultural tradition and economic change are not necessarily antithetical, but the traditions continue to orient the economic change, such that the continuity of tradition consisted in the specific way a people were prepared to change (Hannerz, 1992; Thomas, 1991; Sahlins, 1993).

A report of the World Commission on Culture and Development established by UNESCO reflects the shift in thinking about the concept of development (Pérez de Cuéllar, 1995).

> According to one view, development is a process of economic growth . . . According to the other, espoused by the UNDP's [United Nations Development Programme's] annual *Human Development Report* and by many distinguished economists, development is seen as a process that enhances the effective freedom of the people involved to pursue whatever they have reason to value . . . Poverty of life, in this view, implies not only lack of essential goods and services, but also a lack of opportunities to choose a fuller, more satisfying, more valuable and valued existence. The choice can also be for a different style of development, a different path, based on different values from those of highest income countries now. The recent spread of democratic institutions, of market choices, of participatory management of firms, has enabled different groups and different cultures to choose for themselves.

It can no longer be supposed, as it was in an earlier generation of "modernization" theories, that the non-Western people's culture has something the matter with it. If any general lesson can be drawn from the mixed practical efforts of earlier decades to remake the world in the Western image of progress, it is that a people's economy is part of their larger organization of life, and the culture that supplies its principal relationships and values. As stated by the World Commission on Culture and Development: "Culture is not a means to material progress: it is the end and aim of 'development' seen as the flourishing of human existence in all its forms and as a whole" (Pérez de Cuéllar, 1995). In other words, development in the late 20th century is realized as the indigenization of modernity— the encompassment of modern technologies in local cultures.

The NRC Committee to Review the Community Development Quota found this concept of development particularly appropriate to its investigation. It is useful and appropriate because a working relation between the modern market economy and the existing cultural traditions is already in effect in western Alaskan communities, the market now providing important monetary means for the realization of the cultural traditions. Indeed, the most serious problem facing these communities seems not so much how to synthesize the commercial economy with their own cultural values, as how to get enough income from the market to support the traditional cultural values that are now dependent on it. For the so-called subsistence lifestyle widely practiced in western Alaska by non-native and native people alike is both less and more than the term "subsistence" tends to imply. It is less because subsistence production would be impossible without engagements in the commercial relations and civic institutions of the larger American society, including the income people need for their means of subsistence production. The necessary adoption of modern technology makes money their own cultural capital. But then, "subsistence" means more than it seems since, as an *Anchorage Daily News* reporter put it, it is "not a lifestyle but a life" (Anchorage Daily News, 1989).

A total world order of cultural values, practices and relationships is linked to the annual cycle of harvesting wild animal and plant resources. "Subsistence" includes: (1) the relationships of households, extended families, and larger communities constructed through cooperation in production and customs of reciprocal sharing; (2) the division of labor between men and women and the corresponding understanding of their respective competencies; (3) the accumulation of prestige and influence by certain individuals, such as successful hunters or whaling captains, and certain knowledgeable people, such as experienced elders, which constitutes the political contours of the community; (4) the dances, inter-village exchange festivals, and other social celebrations, often integrated in the calendar of Christian religious and American national holidays; and (5), not least, the distinctive relations of exchange that people understand to exist between themselves and the natural species on which they depend.

Native communities attach very strong values to indigenous foods, a diet of which is considered indispensable to human strength and health. Also associated are certain valued traits of human character, of the kind necessary to undertake an often difficult "subsistence" existence: a very considerable knowledge of nature and a high degree of technical competence (including competence in dealing with modern technologies), highly athletic physical skills, and the sort of mental toughness that combines sagacious prudence with the ability to respond to the emergencies and contingencies of famously difficult Arctic conditions (Nelson, 1969). "Subsistence" is indeed much more than subsistence; it is a whole way of life that extends to the people's essential conceptions of themselves and of the objects of their existence[2] (Jorgensen, 1990).

Yet "subsistence" is no longer a phenomenon of the people's own making, and in that sense we say it is less than a complete existence. It depends decisively and unconditionally on monetary flows from the public and private sectors for the acquisition of necessary capital. There is no going back now to fur clothing, dog sleds, and bone-pointed harpoons; the subsistence economy runs on snow-machines, motorized aluminum fishing vessels, four-wheel all-terrain vehicles, pickup trucks, CB radios, manufactured fishing and hunting gear, fossil fuels,

---

[2] Subsistence also includes the responsibilities that people fulfill in providing for their families and caring for the older people and for other relatives, friends, and neighbors. It includes the pleasure that people derive from subsistence pursuits and from being out in the country. It includes the respect and reverence that people have toward the land and animals. It includes the understanding people have of the deeper connections between themselves as a people, the land and the sea, and the resources. The people know that these connections sustain not only their bodies but also their cultural and personal essences, giving them identity and meaning in their lives as persons and as a people.

camping equipment, imported cold weather clothing, and airplanes (Langdon, 1991). Changes in lifestyle including settlement patterns in the villages, improved safety, the availability of technology, and the desire for other market goods that reduce the time available for subsistence activities have contributed to the increasing importance of capital for conducting subsistence activities. An average of about $20,000 to $25,000 in annual cash expenses per household is a reasonable estimate of what is presently required to maintain an ongoing "subsistence" economy.

As is well known, the Alaska Natives have adopted a great variety of strategies to acquire the income necessary for daily life—although the available sources have not been sufficient to prevent serious and widespread poverty. In broad terms, the principal sources of cash income have been commercial fishing and hunting; craft production; dividends from native corporations; income from participation in the National Guard and state construction projects[3]; loans from government agencies and fiscal institutions; and a variety of transfer payments ranging from Aid to Families with Dependent Children (AFDC) to annual distributions from the Alaska Permanent Fund Dividend. Considering that these flows go largely to maintaining the so-called subsistence system, it follows that the "development" of western Alaskan communities is, in an unusual double sense, an enrichment of their way of life.

It is important to stress the seeming paradoxes of this development, precisely because in western Alaska the market economy and the traditional lifestyles have proved to be compatible despite the conventional social science wisdom, now outmoded, according to which synthesis of money and culture would be impossible. For many years it was believed that money destroys the traditional community, because money *becomes* the community, the effective nexus of relationships between people. But decades of academic studies in western Alaskan communities by a variety of anthropologists and sociologists have repeatedly testified to the contrary; the Alaska Native communities have proven open to technological innovation and adaptable to market economy as well as to government bureaucracy, while all the time integrating these external influences in their own cultural purposes (Hensel, 1996; Nowack, 1975; Van Stone, 1960, 1962). General observations to this effect have been made in villages in the CDQ area. For example, a study conducted on the lower Yukon Delta notes:

> The interaction of monetary income and subsistence output argues against a conceptual polarization of a "subsistence economy" and "market economy" as mutually exclusive and antagonistic production systems. Rather than being competing systems, Kotlik fishermen have blended them in a mutually supportive mixed economy (Wolfe, 1986).

---

[3]Income from the National Guard and state construction projects have fallen in recent years, contributing to a shrinking economic base.

Anne Fienup-Riordan found similar conditions on Nelson Island. She writes:

> The fact that the same men and women who affirm their cultural heritage through the defense of traditional subsistence resource utilization patterns are often the very ones who are anxious to develop the [commercial] economic opportunities available in their villages does not represent a contradiction or a basic incompatibility between cultural values and economic reality. Rather, choices in the realm of economic, political, and social activity are made in accordance with a common cultural value system (Fienup-Riordan, 1986a).

Finally, a study of two Yup'ik communities, Togiak and Quinhagak, in the Bristol Bay and Kuskokwim regions makes a telling observation about the positive relationship between success in the commercial and traditional economies— a finding that has been replicated in a number of Alaska Native villages:

> It is . . . noteworthy, as found elsewhere in rural western Alaska, that "success" in petty commodity production is highly correlated with success in subsistence . . . [T]he high commodity producers choose to be high producers in the subsistence sector as well. They have sufficient time and well organized managerial skills to pursue subsistence. Subsistence activities continue to provide them with satisfaction, status and allow them to fulfill obligations to kinsmen and community. Subsistence is deeply embedded in what it means to be a Yup'ik in these communities (Langdon, 1991).

From this complementary relation of the subsistence and monetary economies follows the finding—extremely relevant for assessing the CDQ program— that higher levels of household cash income are directly correlated with the people's commitment to and their returns from natural resource harvesting (hunting, fishing, and gathering). So again, an observation on Kotlik villagers:

> The members of families with larger monetary incomes had a larger quantity of subsistence food products in their diets than families with smaller monetary incomes. The relationship appears linear and significant at $p = .05$ . . . The implication is that income and subsistence use increase together (Wolfe, 1986).

Indeed, where commercial fishing is involved, the relationship of the two sectors is mutually reinforcing: successful fishing yields increased cash, thus improved subsistence gear, hence greater returns from each subsistence attempt. Yet, interestingly enough, studies show the same positive relationship held where employment is concerned: that men who are able to obtain employment locally, even for the greater part of the year, are at least as committed to subsistence activities as those employed, and are generally more successful. Kruse (1986), for example, found this to be the case in North Slope Iñupiat villages—where the data indicate that subsistence activity increases with the number of months of employment annually, up to those working for wages nine months or more—and the general correlation is also reported for villages in the CDQ program area

(Langdon, 1986, p.32, Wolfe, 1980, 1986; Lonner, 1986). Of course, it is the increased mobility and efficiency provided by modern technology, provided by cash income, that makes it possible for individuals to succeed in both sectors of the economy at once. Extended families, as well as households at a stage in their development cycle that affords them working members in two generations, are even better endowed to exploit simultaneously the monetary and subsistence spheres. In such cases a division of labor—such as between elders and younger members of the family or between young men of different skills and dispositions—will provide the advantage of at least some specialization in the two sectors (Worl and Smythe, 1986). Family size, solidarity, and stage of development are also critical in determining whether outside employment and educational opportunities can be exploited—such as the opportunities expected from CDQ programs.

Another unusual empirical finding deserves emphasis, namely that adults in rural villages who have the greatest "outside" experience in terms of education and employment, which is to say the more "acculturated" or "Westernized" individuals, usually have greater interest and output in the subsistence economy than people who have not had such backgrounds (Kruse, 1986). This relationship is again contingent on higher monetary income, thus improved means of subsistence production, among those who have passed time in urban environments of Alaska or elsewhere. Still, it would come as some surprise to earlier theories of "modernization" that the more Westernized people become, the more indigenous they will be:

> It appears, then, that men remain interested in subsistence activities despite exposure to western influences. In fact, men who for reasons of choice or fate have been exposed to Western influences tend to be *more* interested in subsistence (Kruse, 1986).

For some time now, especially since World War II, the communities of western Alaska have extended beyond the boundaries of their villages. These communities include people working or studying in metropolitan centers of Alaska and the contiguous United States, people who may well return in the classic pattern of "circular migration," but who remain linked to their villages of origin by sentiment, kinship, and the exchange of money, goods, and news. Such linkages have, of course, been facilitated by modern means of transportation and communication. As in many other areas of the world where people, ideas, and commodities are on the move between rural homelands and urban centers of employment, new forms of dispersed communities have developed in the Twentieth Century—and may well be with us for generations to come. The geographic village is small, but the social village extends for thousands of miles. Moreover, the flow of goods—so-called remittances that are viewed locally as part of a traditional reciprocity or sharing system—generally favors the village, hence reverses the historic parasitic economic relationship between cities and rural areas.

It can be expected that Alaska Native communities will have members working in the cities but contributing to the village economy, even if they become long-time city residents. For this reason, varied educational and training programs, even without seeming applicability to rural economic activity, can have great development benefits for the native villages.

The pattern of benefits from education, training, and outside employment, accruing to families (or extended families) in the CDQ villages depends on the demography and the stage in the development cycle of the family. Some households, particularly those composed of young married couples with dependent children and those composed of older community members without access to young people's labor, will not be able to take advantage of economic opportunities outside the village. These people can be served by development projects that emphasize increased economic activities in the villages or nearby rural areas, rather than in distant fishing grounds or urban areas. Village-based development projects include loans for expanding local commercial fishing operations and creation of local wage-earning and marketing opportunities in fish processing and distribution.

Indeed, the circular migration between rural villages and cities is an aspect of recent Alaskan history that affords a ready-made experimental demonstration of just what "development" is in this cultural context—and thus, what might be hoped for from the CDQ program. We mean the cultural "renaissance" as it has been called and as it has been reported from several locations within the CDQ region, coming on the tide of increased cash flows in rural Alaska since the early 1970s. "It was not anticipated," Joseph Jorgensen wrote in a recent work on *Oil Age Eskimos* (1990), "that ANCSA's [Alaska Native Claims Settlement Act] provisions would lure natives back to their natal villages from urban areas in Alaska and elsewhere, but the prospects of employment, land, and money have had that effect." The result of a multi-year investigation by a team of ethnographers, Jorgensen's report included two villages, Unalakleet (Yukon) and Gambell (St. Lawrence Island), in the CDQ program—as well as a third, Wainwright, on the North Slope (Jorgensen, 1990). The influx of money and increase of employment opportunities in these communities came from activities of native corporations, government projects, and the North Slope oil development, as well as enhanced commercial fishing and craft production. Federal housing projects in the villages, unrelated to ANSCA, have also contributed by providing an incentive to lure people back to their villages. Immigration, as well as efforts to eradicate diseases in these communities, also contributed to population increase. Gambell, for example, grew from 372 in 1970 to 522 in 1989.

If the population increase was not anticipated, even less expected was the "renaissance of native culture" that accompanied it—what the researchers characterized summarily as a resurgence of "subsistence pursuits, native lore and legends, native dancing and singing" (Jorgensen, 1990). These were not the only aspects of the local culture that found new life. Relations of kinship were extended and (on St. Lawrence Island) clanship strengthened, as these were engaged in the organization of production and exchange. It is "an interesting fact that should not go unnoticed," observed Jorgensen (1990), "that the more people earned, whether in private sector jobs or as successful private sector fisherman, the more widely they shared." On Nelson Island (also in the CDQ program), a similar cultural efflorescence has been described by Anne Fienup-Riordan (1983, 1986, 1990). The focus has been the traditional springtime seal festival, celebrating the first catch of the season, the occasion of elaborate distributions of meat and goods that realize the main social relationships of these Yup'ik communities. The feast and exchanges, including inter-village exchanges, have seen "an incredible quantitative elaboration" in recent times (Fienup-Riordan, 1986). Noting that the forecasts of a waning subsistence lifestyle so popular in the 1950s and 1960s have not been confirmed, Jorgensen sums up such surprising development effects as follows:

> To the contrary, political and economic events since 1970 have had the contradictory consequences of causing Alaskan Eskimos to become increasingly dependent on the public and private sectors of the national economy but also to hunt, fish, and collect more efficiently. Furthermore, the economic and political forces of the past fifteen years have triggered a renaissance of Eskimo dancing and singing, a return migration to villages from urban areas . . . In short, there is a determination on the part of Eskimos to maintain traditional Eskimo culture and at the same time to adapt a pragmatic acceptance of the benefits of modern technology (1990).

What will happen if the CDQ program is economically successful, if it brings increased employment opportunities and moneys to western Alaskan communities, is not entirely unpredictable. The predictions are not the same as might have been made by social scientists (and others) 20 or 30 years ago, but then, they are based on the unexpected experience of recent decades, which have seen the "development" of traditional culture in places where access to the market economy, monetary incomes, and technological means of subsistence production have all improved. "Development" thus indeed turns out to be "the enrichment of a way of life." But of course it would be very wrong to suppose that this process is ever economically simple or culturally unproblematic. "Enrichment" is never easy, nor are all modes of doing so supportive of the local way of life—or the people's sense of self-respect.

It is a popular opinion that the social problems of indigenous Alaskan communities—alcoholism, drug abuse, and suicide most notably—are the results of a "clash of cultures." Presumably the unfulfillable expectations and intractable barriers set up by American society, together with the declining strength and appeal of native custom, have been the principal ingredients of a widespread despair. On the other hand, the information we reviewed suggests that the explanations of despondency by reason of cultural conflicts are not altogether sufficient. The great success stories in native communities figure as their main protagonists the people who have had the most "outside" experience and ability in the market economy—success consists precisely in turning these assets to the enhancement of the "subsistence lifestyle." The leaders of some native communities in western Alaska are often among the most "acculturated" of the local people. It is their competence in both cultures, the native and the Western, that allows them to synthesize the differences—and gives them also a measure of local respect. Hence the problems of modern native communities do not necessarily reside in inherent cultural incompatibilities so much as in situations that make it difficult for people to synthesize the differences.

If development is "the enrichment of the community's way of life," it is important to stress again that, as the situation now stands in western Alaska, the traditional way of life cannot proceed without cash. In this respect, young men and women and young families are in a particularly precarious position. Many of the skills that traditionally equipped people for an honorable and satisfying existence—such as, for men, knowledge of nature, hunting skills, dog sledding, kayaking, whaling—have been rendered technologically obsolete and lost to the younger generation. Unless young people can acquire a monetary stake to subsidize their customary productive activities with the technologies now required, they are in danger of becoming a lost generation. The situation is all the more critical because of the role of autonomy in traditional cultures, that is, on the ability to provide for oneself and family, and beyond that to achieve community standing by supplying others, especially elders and poorer people, with shares from successful subsistence endeavors. As we noted earlier, the problem is not so much that money and the traditional way of life are incompatible, but that without money one cannot participate in the traditional way of life. In that event, the failure is compounded: it is a failure in both cultures. Unable to function in their own society, left with nothing to do and no possible future, the young people are also left out of Euroamerican culture, the good things of which are in their view but not within their grasp. It is a formula for despondency.

As matters have evolved, it is necessary for western Alaskans to succeed, one way or another, in the Euroamerican culture in order to find a place, and peace, in the native culture. If the CDQ program is to have serious developmental consequences, it will have to open the possibilities, especially for young people, to make a go of it in their local communities. But at the same time, there is something more to the value of autonomy that engages the villages and regional organizations as such, the structures by which the CDQ program is constituted. More than any previous welfare or development initiative, more even than the native corporations, the CDQ program seems to offer a viable way for local people to gain control over the means by which they are articulated to the larger economy and society.

This would not only be true of the development councils set up by the CDQ groups but also of the educational training grants they provide. In this regard, it should be remarked that "development" has come to include—not only in Native Alaska, but generally in the world—a growing aspiration for self-determination, both as the means and the end. The people seek a measure of governance, one that will allow them to shape their own future—not only in ways that safeguard their language, values, and customs but according to these cultural desiderata. Such distinctive control by and for members of local communities has thus become a crucial condition of development. Many of the hopes for the CDQ program that NRC committee members encountered in their site visits came from this promise of self-determination—by invidious contrast to welfare handouts and other projects that wind up confirming people in their dependency without relieving their despondency.

3

Overview of the Community Development Quota Program

The Community Development Quota (CDQ) program was implemented in 1992 and has made considerable resources from the Bering Sea fishery available to rural communities in western Alaska. The program allocates a set quota to participating communities and requires those communities to use any earnings to further economic development in the community through investments in fisheries-related industries, infrastructure, and education. The 56 communities initially certified as eligible organized themselves into 6 community groups called CDQ groups. The CDQ groups then formed partnerships with established corporations to participate in the highly industrialized Bering Sea fishery. Benefits generated have included direct revenues from the fishery as well as employment and increased opportunities for the development of fishing infrastructure. The CDQ program began with pollock, was followed by halibut and sablefish, and has since expanded to include crab and a combination of groundfish species, including Pacific cod and other flatfish known as the "Bering Sea multispecies complex."

ORIGIN

The CDQ program emerged as a concept in the mid-1980s, when the Bering Sea groundfishery was transferred from the foreign to the domestic fleet. The new fishery was very profitable for the domestic fishermen. However, one segment of the community received little benefit: the Alaska Native fishermen of rural western Alaskan, who generally came from small villages needing economic development.

The CDQ program was proposed during Congressional hearings on the 1990 reauthorization of the Magnuson Fishery Conservation and Management Act as a way of allowing rural western Alaska access to the fishery. However, the program was not included in the final version of the bill. Discussions about a community development program continued in the North Pacific Fishery Management Council (NPFMC) during debates on two related issues: the split of the allocation of pollock quota between inshore and offshore processing operations and the proposed individual fishing quota (IFQ) program for the halibut and sablefish fishery.

The allocation of pollock quota between inshore and offshore processing operations is central to the management of the North Pacific groundfish fishery. Currently, a percentage of the pollock quota is allocated to inshore, shorebased processing plants that harvest the pollock with their own vessels, or purchase their product from catcher vessels operating in the BSAI pollock fishery. The remainder of the harvest is currently allocated to an "offshore sector" consisting of catcher-processors that harvest and process the pollock on board, processing motherships that receive pollock for processing at sea from catcher vessels, and catcher vessels that deliver pollock to catcher-processors. The allocations between the inshore and offshore sectors has been contentious and a focus of considerable debate over the years. Likewise, the decision to create an individual fishing quota program for the halibut and sablefish fishery was a contentious and difficult decision. In some cases, the CDQ program was linked to negotiations over the inshore and offshore allocations and the consideration of the Alaskan halibut and sablefish IFQ program.

Discussions continued over the next few months and in April 1991 the Council adopted a specific CDQ plan for analysis by council staff and inclusion in the inshore-offshore allocation proposal. The final inshore-offshore proposal (Amendment 18 to the Bering Sea and Aleutian Island Fishery Management Plan) included as a subpart the establishment of a CDQ program in western Alaska and allocated 7.5 percent of the total allowable catch (TAC), or one-half of the reserve quota, to communities on the Bering Sea coast. The decision to allocate 7.5 percent of the TAC (one-half the reserve quota) appears to be a decision based on a combination of factors such as stock assessment, management goals, and political factors. It is not based on specific biological, economic, or social criteria. The 15 percent reserve quota for the TAC is also somewhat arbitrarily defined, but is based on expectations of what percentage of quota might be appropriate to serve as a buffer to prevent overfishing. The allocations to the other species later included in the CDQ program were derived through similar decision-making processes. The plan was approved by the Council in June 1991 and was forwarded to the Secretary of Commerce. On May 28, 1992, the Secretary of Commerce approved the amendment and promulgated implementing regulations (Box 3.1).

In December 1992, state and federal officials finalized the regulations and procedures for the CDQ program. (See Appendixes D and E for state and federal regulations.) In 1995, sablefish and halibut were included in the CDQ program.

BOX 3.1 Summary of Key CDQ Program Elements

A western Alaskan community is eligible for quota if the community:
- is within 50 miles of the Bering Sea;
- is an Alaska Native Claims community;
- has residents who conduct 50 percent of their commercial or subsistence fishing activity in the Bering Sea; and
- did not already have significant pollock activity.

The 56 communities* eligible under this definition organized themselves into 6 groups of varying sizes:
- Aleutian Pribilof Island Community Development Association (APICDA);
- Bristol Bay Economic Development Corporation (Bristol Bay or BBEDC);
- Central Bering Sea Fishermen's Association (Central Bering Sea or CBSFA);
- Coastal Villages Fishing Cooperative (Coastal Villages or CVFC);
- Norton Sound Economic Development Corporation (Norton Sound or NSEDC); and
- Yukon Delta Fisheries Development Association (Yukon Delta or YDFDA).

* Akutan was added as the 57th community in 1996.

Also in 1995, the Council approved a three-year extension of the pollock program and the inshore-offshore allocation. The Council also approved the allocation of quota shares of other groundfish and crab to the CDQ communities in 1995[1]. In the reauthorization of the Magnuson–Stevens Fishery Management and Conservation Act in 1996, the CDQ program was made a permanent program.

---

[1]The history of the multispecies CDQ program began in 1995 when the North Pacific Council approved the License Limitation Program in the groundfish and crab fisheries to prevent new participants from entering the fishery. As part of the License Limitation Program, the council recommended that all other fisheries resources managed under the Bering Sea Aleutian Island management plan be allocated to a CDQ program. However, due to the complexity of the program and other factors, a proposed rule for the multispecies CDQ was not published by the National Marine Fisheries Service (NMFS) until 1997. At this time, the crab CDQ, which had been a part of the multispecies groundfish CDQ was removed from the multispecies framework and harvesting in the crab CDQ fishery began in March 1998. A final rule for implementing the multispecies CDQ program was approved by NMFS in June 1998, and groundfish harvesting is scheduled to begin in October 1998. Some of the complexities that delayed the implementation of the multispecies CDQ are the difficulty in managing the complex quota allocations of the species, as well as the stringent requirements for the program including weighing catch at-sea, data reporting, and on-board observer requirements.

CDQ MANAGEMENT STRUCTURE

Participation in the CDQ program requires an understanding of the complex institutional array that controls management of commercial fisheries in the Bering Sea and Aleutian Islands (BSAI). The six primary components of this institutional structure are:

- the Magnuson-Stevens Fishery Conservation and Management Act;
- the North Pacific Fishery Management Council;
- the National Marine Fisheries Service;
- the Alaska Board of Fisheries;
- the Alaska Department of Fish and Game; and
- the International Pacific Halibut Commission.

Under the Fishery Conservation and Management Act of 1976 (FCMA, renamed the Magnuson-Stevens Fishery Conservation and Management Act in 1996), Congress established a fishery conservation zone extending from 3 to 200 nautical miles offshore. This fishery zone was subsequently incorporated into the U.S. Exclusive Economic Zone (EEZ) established by presidential proclamation. For fisheries inside of three miles, jurisdiction lies with the relevant coastal states. For fisheries in the EEZ, the Magnuson-Stevens Fishery Conservation and Management Act (MSFCMA) created a new management system consisting of eight regional fishery management councils, with the relevant council for fisheries off Alaska being the North Pacific Fishery Management Council.[2]

Under the MSFCMA, the NPFMC has limited authority over the fisheries of the Arctic Ocean, Bering Sea, and Pacific Ocean in the EEZ off Alaska. The nature of this limited authority is that Congress charged the Council with development of fishery management plans, and subsequent amendments, for these fisheries when necessitated by conservation and management concerns. Once the Council develops a plan, it is then submitted to the U.S. Secretary of Commerce for review and approval or disapproval. Typically, the councils, NMFS staff, and the National Oceanic and Atmospheric Administration (NOAA) general counsel coordinate with each other in developing fishery management plans and regulations. For approved plans, the U.S. Secretary of Commerce (acting through the National Marine Fisheries Service) is charged to develop implementing regulations and enforce those regulations.

The CDQ program falls under two of the fishery management plans developed by the Council, the groundfish fishery in the BSAI area and the commercial king and Tanner crab fisheries. Council actions are intended to represent a delib-

---

[2]The other regional fishery management councils are the New England Council, the Mid-Atlantic Council, the South Atlantic Council, the Caribbean Council, the Gulf of Mexico Council, the Pacific Council, and the Western Pacific Council.

erative policy process but these actions must also comply with an array of legal requirements. All fishery management plans and amendments devised by the Council must comply with the MSFCMA. In particular, the MSFCMA prescribes national standards for fishery management measures that must be adhered to when designing and selecting management measures. In addition to the MSFCMA, Council actions must also comply with other applicable law.

To further unravel the institutional matrix relevant to the CDQ program, it is useful to consider the conventional separation of the commercial fisheries of the BSAI into three groups: groundfish, crab, and halibut. In the vernacular of the Council "groundfish" refers to most of the commercially valued species of fish except salmon, herring, and halibut. Groundfish fisheries in the BSAI are managed by the Council and the National Marine Fisheries Service. The Council sets TAC levels for various target species fisheries that occur throughout the EEZ and state waters. The Scientific and Statistical Committee assists the Council by providing advice on the level of the allowable biological catch (ABC), from which the council sets a TAC lower than the ABC. State waters are effectively included in the federal BSAI groundfish management regime through a process of concurrent opening and closing of state waters keyed to federal openings and closings.

For the commercial crab fisheries in the BSAI, the principal regulating institutions are the Alaska Board of Fisheries, the Alaska Department of Fish and Game (ADF&G), and for specific measures, the Council. The Board of Fisheries and ADF&G manage the commercial crab fisheries of the BSAI in both state and federal waters. The Alaska Board of Fisheries has regulatory authority for fisheries within Alaska, and ADF&G provides technical, scientific, and managerial support to the Board of Fisheries. State management of the crab fisheries in federal waters occurs under authority delegated to the state by the Council. Notably, the Council retained authority over limited access and other forms of specific allocations (such as a crab CDQ allocation) in the BSAI crab fisheries. The MSFCMA and other applicable federal laws control both Council actions and actions pursued under authority delegated to the state by the Council.

The International Pacific Halibut Commission (IPHC) manages the commercial halibut fishery in the BSAI under authority provided by international treaty (the "Convention Between the United States and Canada for the Preservation of the Halibut Fishery of the Northern Pacific Ocean and the Bering Sea") and the Northern Pacific Halibut Act of 1982 (Halibut Act). The IPHC establishes seasons, determines TAC levels, and apportions the TAC between the United States and Canada. Under the Halibut Act, the Council is authorized to allocate the U.S. TAC established by the IPHC including the use of limited access regimes, such as the CDQ program.

The structure of the Council itself adds to the institutional matrix relevant to commercial fisheries in the BSAI. For the Council, Congress prescribed eleven voting members: six from Alaska, three from Washington, one from Oregon, and one representing the National Marine Fisheries Service. Congress also provided that the Council receive advisory input from a Scientific and Statistical Committee and an Industry Advisory Panel.

ELIGIBLE CDQ COMMUNITIES

To participate, CDQ communities must be within 50 miles of the Bering Sea coast from the Bering Strait to the westernmost Aleutian island or located on islands in the Bering Sea. The communities must be certified as villages that meet the requirements of the Alaska Native Claims Settlement Act (ANCSA). The residents of the community must conduct more than half of their current commercial or subsistence fishing effort in the waters of the Bering Sea and Aleutian Islands. In addition, the community must not have developed previously a capability sufficient to support substantial fisheries, unless the community can show that CDQ benefits would be the only way to realize a return on previous investments. (The community of Unalaska was excluded under this provision.)

Fifty-six eligible communities were originally approved and the community of Akutan was added to the list of approved communities in 1996, when it demonstrated that the local community was not deriving significant benefit from the existing fishing infrastructure. These communities organized themselves into six development corporations, or CDQ groups, each with corporate fishing partners (Table 3.1). Table 3.2 provides an overview of population and economic statistics for these communities. In many of these communities, the levels of poverty and unemployment are significant. In some cases, measures of employment can be difficult to determine because of the limited employment opportunities that exist. This can lead to some inconsistencies when using traditional measures of employment, and the actual employment in these communities may be higher. The majority of eligible communities are small rural villages. Although it is not a requirement that only Alaska Native residents derive benefit from the program, with few exceptions the population of these communities is predominantly Alaska Native, and all of the villages are certified as an Alaska Native Claims community under the Alaska Native Claims Settlement Act. In most cases, the poverty and unemployment levels are high in these villages.

DESCRIPTION OF THE COMMUNITY DEVELOPMENT GROUPS

The Aleutian Pribilof Island Community Development Association

The Aleutian Pribilof Island Community Development Association (APICDA) represents the villages of Akutan, Atka, False Pass, Nelson Lagoon, Nikolski, and St. George. Approximately 400 people live in these 6 villages. In 1998, APICDA received quota allocations of 16 percent of the quota for Bering Sea CDQ pollock, 10 percent of the CDQ quota for Aleutian Islands subarea CDQ sablefish, and various percentages of crab and other groundfish species

TABLE 3.1   Community Development Quota Groups and Pollock Harvesting Partners in 1998

Groups	Partner
Aleutian Pribilof Island Community Development Association (APICDA)	Trident Seafoods Corporation Starbound Partnership
Bristol Bay Economic Development Corporation (BBEDC)	Arctic Storm, Inc.
Central Bering Sea Fishermen's Association (CBSFA)	American Seafoods Company
Coastal Villages Fishing Cooperative (CVFC) Coastal Villages Region Fund (CVRF)	Tyson Seafoods and Westward Seafoods
Norton Sound Economic Development Corporation (NSEDC)	Glacier Fish Company, Ltd.
Yukon Delta Fisheries Development Association (YDFDA)	Golden Alaska Seafoods, Inc.

NOTE: Fishing partners may change over time.

(Tables 3.3, 3.4, and 3.5). The Atka Fishermen's Association, a group of halibut fishermen within the APICDA communities, harvests 100 percent of the CDQ allocation of halibut in the western Aleutian region (area 4B). APICDA has formed partnerships with Trident Seafoods Corporation, a processing company with operations throughout Alaska, and Starbound Partnership, the operator of the F/V *Starbound* for harvesting its pollock quota. APICDA jointly purchased a factory longliner, the F/V *Rebecca B*, with the Yukon Delta Fisheries Development Association for harvesting the halibut and sablefish quota. Since that time, the *Rebecca B* has been grounded, declared unsalvagable, and sunk. Additional details on the investments pursued by APICDA are provided in Appendix F.

Bristol Bay Economic Development Corporation

The Bristol Bay Economic Development Corporation (BBEDC) represents the villages of Aleknagik, Clark's Point, Dillingham, Egegik, Ekuk, King Salmon/Savonoski, Manokotak, Naknek, Pilot Point, Port Heiden, South Naknek, Togiak, Twin Hills, and Ugashik. Approximately 5,300 people live in these 14 villages. The quota allocations for 1998 include 20 percent of the Bering Sea CDQ pollock, 25 percent of the Aleutian Islands subarea CDQ sablefish, 23 percent of the

TABLE 3.2 Population and Economic Statistics of Approved Communities in CDQ Program (information from DCRA database)

Community	Population 1997	% Native Population	Median Value/ Home ($)	Total Households (Occupied)	Persons/ Household	Median Household Income ($)	% Unemployment	% Poverty
YDFDC								
Alakanuk	651	95.80	34,200	121	5	17,708	26.80	29.40
Emmonak	820	92.10	46,300	161	4	25,625	34.60	20.90
Kotlik	543	97.00	40,600	101	5	20,417	36.60	17.70
Sheldon Point	177	92.70	14,999	27	4	16,250	13.00	56.20
NSEDC								
Brevig Mission	265	92.40	101,700	53	4	15,000	35.30	24.70
Diomede	174	93.80	14,999	41	4	14,375	0.00	63.00
Elim	301	91.70	61,100	73	4	16,250	36.10	25.10
Gambell	653	96.20	39,400	120	4	15,938	16.80	46.40
Golovin	152	92.90	62,500	42	3	16,146	15.30	8.30
Koyuk	272	94.80	63,100	61	4	18,750	37.30	30.00
Nome	3,656	52.10	79,900	1,119	3	45,812	11.00	9.90
St. Michael	341	91.20	27,500	69	4	23,194	22.90	20.90
Savoonga	622	95.20	49,600	116	4	11,339	14.70	50.90
Shaktoolik	226	94.40	28,500	46	4	18,438	31.90	22.80
Stebbins	513	94.80	21,700	86	5	23,333	39.20	32.30
Teller	265	86.80	22,500	68	3	20,000	3.30	32.10
Unalakleet	803	81.80	65,000	207	3	34,531	19.20	11.60
Hooper Bay	1,012	96.00	24,300	190	4	18,125	41.70	43.50
Kipnuk	567	97.40	31,400	99	5	4,999	12.90	76.60
Kongiganak	349	97.30	14,999	60	5	33,250	16.30	30.30
Kwigillingok	333	95.00	17,300	62	4	14,500	9.20	43.10
Mekoryuk	192	99.40	76,400	63	3	14,792	16.70	31.50
Hooper Bay	1,012	96.00	24,300	190	4	18,125	41.70	43.50
Kipnuk	567	97.40	31,400	99	5	4,999	12.90	76.60

Kongiganak	349	97.30	14,999	60	5	33,250	16.30	30.30
Kwigillingok	333	95.00	17,300	62	4	14,500	9.20	43.10
Mekoryuk	192	99.40	76,400	63	3	14,792	16.70	31.50
Newtok	269	93.20	23,800	42	5	14,844	25.90	50.20
Nightmute	217	95.40	41,300	29	5	17,813	26.90	62.00
Platinum	41	92.20	14,999	22	3	23,056	8.00	35.80
Quinhagak	567	93.80	60,500	127	4	17,500	5.90	37.20
Scammon Bay	459	96.50	76,600	85	4	15,179	18.40	40.70
Toksook Bay	496	95.50	105,400	88	5	21,875	25.50	39.20
Tuntutuliak	351	96.70	28,300	70	4	14,444	6.40	46.00
Tununak	330	96.20	42,500	78	4	18,750	14.00	26.30
BBEDC								
Aleknagik	226	83.20	112,500	57	3	21,875	14.30	28.80
Clark's Point	66	88.30	87,500	18	3	17,083	18.50	16.10
Dillingham	2,252	55.80	106,100	691	3	44,083	6.70	9.50
Egegik	127	70.50	33,800	48	3	20,625	24.30	34.10
Ekuk	2	33.30	0	1	3	0	0.00	0.00
King Salmon	478	15.50	143,800	158	3	54,072	5.80	3.00
Manokotak	387	95.60	41,300	90	3	20,500	16.10	28.60
Naknek	640	41.00	108,900	208	3	50,907	3.90	1.70
Pilot Point	115	84.90	93,800	17	3	38,750	0.00	12.90
Port Heiden	116	72.30	101,000	42	3	35,000	22.00	24.30
South Naknek	149	79.40	54,600	39	3	23,750	27.50	26.30
Togiak	762	87.30	32,500	151	4	15,000	23.10	46.30
Twin Hills	59	92.40	20,000	25	3	11,667	25.00	50.00
Ugashik	5	85.70	0	4	0	37,500	80.00	0.00
CBSFA								
St. Paul	64	66.10	84,100	154	4	39,922	10.80	7.10

SOURCE: Alaska Department of Community and Regional Affairs, 1998.

TABLE 3.3   Percent Allocations of the 7.5 percent TAC of Pollock

	1992-93	1994-95	1996-98
APICDA	18	18	16
BBEDC	20	20	20
CBSFA	10	8	4
CVFC/CVRF	27	27	25
NSEDC	20	20	22
YDFDA	5	7	13
TOTALS	100	100	100

SOURCE: Alaska Department of Community and Rural Affairs, 1998

TABLE 3.4   1997 Percent Allocations of the TAC for Pollock, Sablefish, and Halibut

	APICDA	BBEDC	CBSFA	CVFC/ CVRF	NSEDC	YDFDA	TOTAL
Pollock	16	20	4	25	22	13	
Sablefish,							
Hook & Line-AI	10	25		25	30	10	
Sablefish							
Hook & Line-BS					25	75	
Halibut							
Area 4B	100[a]						
Area 4C				100			
Area 4D		23		24	25	33	
Area 4E		30		70			

SOURCE: Alaska Department of Community and Regional Affairs, 1998.
[a]Atka Fisherman's Association

halibut in the western Bering Sea region (area 4D), 30 percent of the halibut in the eastern Bering Sea region (Area 4E), and various percentages of crab and other groundfish species (Tables 3.3, 3.4, and 3.5). BBEDC has formed partnerships with Arctic Storm, Inc., to harvest the pollock quota and the F/V *Recovery* for harvesting the halibut in the western Bering Sea (area 4D). BBEDC has also formed a partnership with Alaskan Leader, Inc., for harvesting part of their multispecies allocation and for the joint ownership of a freezer longliner, F/V

TABLE 3.5   1998-2000 Percent Allocations of the TAC for Crab, Multispecies Groundfish, Bycatch, and Prohibited Species

	APICDA	BBEDC	CBSFA	CVFC/ CVRF	NSEDC	YDFDA	TOTAL
Halibut							
4B	100	0	0	0	0	0	100
4C	10	0	90	0	0	0	100
4D	0	23	0	24	26	27	100
4E0	30	0	70	0	0	100	
Crab							
Bristol Bay red king	20	20	0	20	20	20	100
Norton Sound red king	0	0	0	0	50	50	100
Pribilof red & blue king	0	0	100	0	0	0	100
St. Matthew blue king	50	12	0	12	14	12	100
Bering Sea C. opilio							
Tanner	10	19	19	17	18	17	100
Bering Sea C. bairdi							
Tanner	10	19	19	17	18	17	100
Sablefish & Turbot							
Sablefish,							
hook & line-AI	15	20	0	30	20	15	100
Turbot-AI	18	18	5	14	26	19	100
Sablefish							
hook & line-BS	15	22	18	0	20	25	100
Turbot-BS	16	25	14	1	20	24	100
Pacific cod	16	20	10	17	18	19	100
Atka mackerel							
Eastern	20	17	10	17	16	20	100
Central	20	17	10	17	16	20	100
Western	20	17	10	17	16	20	100
Yellowfin Sole	29	25	8	5	5	28	100
Flatfish							
Other flats	20	20	10	15	15	20	100
Rock sole	10	20	10	20	20	20	100
Flathead	20	20	10	15	15	20	100
Squid	19	18	10	17	16	20	100
Other species	19	22	9	14	14	22	100
Other Rockfish							
O. rockfish-BS	16	20	8	18	19	19	100
O. rockfish-AI	16	20	8	18	19	19	100
Arrowtooth	19	21	9	15	15	21	100

TABLE 3.5 cont.

	APICDA	BBEDC	CBSFA	CVFC/ CVRF	NSEDC	YDFDA	TOTAL
Pacific Ocean perch complex							
True POP-BS	20	17	10	17	16	20	100
Other POP-BS	20	17	10	17	16	20	100
True POP-AI:							
Eastern	20	17	10	17	16	20	100
Central	20	17	10	17	16	20	100
Western	20	17	10	17	16	20	100
Sharp/northern-AI	20	17	10	17	16	20	100
Short/rougheye-AI	17	20	9	17	18	19	100
Sablefish, trawl-AI	16	20	10	17	18	19	100
Sablefish, trawl-BS	16	20	10	17	18	19	100
Prohibited Species Quota							
Halibut (mt)	20	22	8	13	14	23	100
Herring (mt)	17	17	16	17	17	16	100
Chinook salmon(#)	21	21	9	13	13	23	100
Other salmon(#)	23	23	8	11	11	24	100
C. bairdi-zone 1(#)	24	25	7	9	9	26	100
C. bairdi-zone 2(#)	24	25	7	9	9	26	100
Red king crab(#)	19	21	9	15	15	21	100

SOURCE: Alaska Department of Community and Regional Affairs, 1998

Bristol Leader. Resident fishermen using local vessels harvest the halibut in the eastern Bering Sea region (Area 4E). Additional details on the investments pursued by BBEDC is provided in Appendix F.

Central Bering Sea Fishermen's Association

The Central Bering Sea Fishermen's Association (CBSFA) represents the village of St. Paul in the Pribilof Islands. St. Paul has approximately 750 residents. The quota allocations for 1998 include 4 percent of the Bering Sea CDQ pollock, and various percentages of crab and other groundfish species (Tables 3.3, 3.4, and 3.5). The Pribilof Island Fishermen's Association, which included the villages of St. Paul and St. George and was a member of APICDA, harvested 100 percent of the halibut in the Pribilof Island region prior to 1997 (Area 4C). The village of St. George now receives its halibut allocation through the APICDA CDQ group. CBSFA has formed a partnership with American Seafoods to harvest the pollock quota. Additional details on the investments pursued by CBSFA are provided in Appendix F.

Coastal Villages Fishing Cooperative/ Coastal Villages Region Fund

The Coastal Villages Fishing Cooperative (CVFC) and the recently formed Coastal Villages Region Fund (CVRF) represent the villages of Chefornak, Chevak, Eek, Goodnews Bay, Kipnuk, Kongignak, Kwigillingok, Mekoryuk, Newtok, Nightmute, Platinum, Quinhagak, Scammon Bay, Toksook Bay, Tuntutuliak, and Tununak. Approximately 5,769 people live in these 17 communities. The quota allocations for 1998 include 25 percent of the Bering Sea CDQ pollock, 25 percent of the Aleutian Islands management subarea of CDQ sablefish, 24 percent of the halibut in the western Bering Sea region (Area 4D), 70 percent of the halibut in the eastern Bering Sea region (Area 4E), and various percentages of crab and other groundfish species (Tables 3.3, 3.4, and 3.5). CVFC/CVRF has formed a partnership with Tysons Seafoods and Westward Seafoods to harvest the pollock quota. Additional data on the investments pursued by CVFC/CVRF are provided in Appendix F.

Norton Sound Economic Development Corporation

The Norton Sound Economic Development Corporation (NSEDC) represents the villages of Brevig Mission, Diomede/Ignaluk, Elim, Gambell, Golovin, Koyuk, Nome, St. Michael, Savoonga, Shaktoolik, Stebbins, Teller, Unalakleet, Wales, and White Mountain. Approximately 7,745 people live in these 15 communities. The quota allocations for 1998 include 22 percent of the Bering Sea CDQ pollock, 30 percent of the Aleutian Island management subarea of CDQ sablefish, 25 percent of the Bering Sea management subarea of CDQ sablefish, 20 percent of the halibut in the western Bering Sea region (area 4D), and various percentages of crab and other groundfish species (Tables 3.3, 3.4, and 3.5). NSEDC has formed a partnership with Glacier Fish Company, Ltd., to harvest the pollock quota, and in October 1997 acquired a 50 percent share in the Glacier Fish Company for approximately $40 million. Additionally, NSEDC and Glacier Fish Company jointly own a freezer longliner, the F/V *Norton Sound,* to harvest the CDQ sablefish quota. The halibut quota is harvested by fishermen from Nome, Savoonga, and other villages using local vessels. Additional data on the investments pursued by NSEDC are provided in Appendix F.

Yukon Delta Fisheries Development Association

The Yukon Delta Fisheries Development Association (YDFDA) represents the villages of Alakanuk, Emmonak, Kotlik, and Sheldon Point. Approximately 1,700 people live in these 4 communities. The quota allocations for 1998 include

13 percent of the Bering Sea CDQ pollock, 33 percent of the halibut in the western Bering Sea region (Area 4D), 10 percent of the Aleutian Islands management subarea of CDQ sablefish, 75 percent of the Bering Sea management subarea CDQ sablefish, and various percentages of crab and other groundfish species (Tables 3.3, 3.4, and 3.5). YDFDA has formed a partnership with Golden Alaska Seafoods, Inc., to harvest the pollock quota. The halibut and sablefish quotas are harvested by local residents using vessels owned by YDFDA. Additional data on the investments provided by YDFDA are provided in Appendix F.

ALLOCATION OF QUOTA

The amount of fish quota allocated to western Alaskan communities is 7.5 percent of the total allowable catch of Bering Sea multispecies fishery, 20 percent of the BSAI sablefish, and various percentages of the halibut in Bering Sea management Areas 4A though 4E. The program was recently expanded to include Pacific cod, Atka mackerel, turbot, yellowfin sole and several other species of flatfish, as well as king, Tanner, and snow crab. Under the multispecies program, CDQ groups will be allocated a total of 7.5 percent of the quota. Bering Sea opilio, bairdi, and king crab will be phased in at 3.5 percent in 1998, 5 percent in 1999, and 7.5 percent in 2000.

The State of Alaska evaluates applications and recommends quota allocation for each CDQ applicant. The U.S. Secretary of Commerce and NPFMC review the state's recommendations and the Secretary of Commerce makes the final authorization for CDQ applicants to harvest quota. The maximum award to any group is 33 percent of the overall 7.5 percent CDQ allocation. The criteria used to allocate quota are addressed in Chapter 4.

In 1994, past performance was included in the criteria for allocations. For the 1996-1998 period, significant adjustments in the allocated quota were made on the basis of past performance. For example, pollock allocations were reduced for central Bering Sea and increased for Norton Sound and Yukon Delta (Table 3.3). Table 3.3 provides greater detail on the full range of reallocations.

PHASES OF CDQ DEVELOPMENT

The primary purpose of the CDQ groups is to determine the best use of the allocations provided to them. This simple purpose requires considerable deliberation because each group must decide which activities are best suited for its region and constituents. Examination of CDQ activities since 1992 indicates that four phases have occurred:

- Phase I: Formation and governance
- Phase II: Organization direction and plan development

- Phase III: Plan implementation
- Phase IV: Performance review and revision

While there have been certain defining elements to each of these phases of development, some activities begun in phase I have necessarily continued through the other phases. Each phase has been characterized by its own distinctive, essentially non-recurring activity as well. This section will identify the range of program activities that have occurred in Phases I-IV.

Phase I: Formation And Governance

Formation

Activities include identification of member communities, structure of governance, by-laws, and, if necessary, incorporation. The type of organization chosen (cooperative, for-profit corporation, nonprofit corporation) often has distinctive influences on the subsequent direction of the group. Many CDQ groups have created a network of subsidiaries (some as for-profit corporations), partnerships, and joint venture arrangements, which they see as best suited to carrying out the different activities.

Governance

This includes setting up a management organization with a decision-making structure and executive leadership, establishing financial oversight capability, establishing recognized ties to appropriate federal and state bureaucracies, and developing acceptable Community Development Programs. It also means setting up legal arrangements for contracts and ensuring the fiduciary responsibility of the CDQ group.

For the existing CDQ groups, Phase I occurred mainly in the summer and early fall of 1992. Each of the eligible communities was required to hold meetings at which fishermen were selected to represent the community (State of Alaska, 1995). Subsequently, six applicant CDQ groups emerged "based primarily on geographical proximity and cultural boundaries" (State of Alaska, 1995, p. 11). To some extent, experience with the corporate model used in ANSCA villages guided the structure of the CDQ groups. There is substantial variability in the groups largely due to the lack of specificity in federal and state regulations about how communities were to organize or implement the quotas. The election procedures used by the CDQ groups are established in bylaws and are similar. All of the groups have established criteria for membership on the board of directors that guides the CDQ group. Most of the groups require that a representative on the board of directors be a permanent resident of the region or village from which they are elected for a certain period of time, and a certified commercial or subsistence fisherman. The representatives are elected to the board of directors through recognized governing bodies in the communities such as the traditional council, Elder's council, City Council, local village corporation, or the Indian Reorganization Act council. Typically, each village is allowed to elect one representative to the board of directors. Representatives on the board of directors typically serve for two or three year terms, although in some cases the terms of office are not specified. Several of the CDQ groups also have term limits for elected representatives.

The board of directors can also elect officers to aid in the administration of the CDQ groups, including a President, Vice President, Secretary and Treasurer. Additionally, most of the CDQ groups have provisions to allow the hiring of an Executive Director who can act on behalf of the Board of Directors. Some of the CDQ groups have provisions allowing the formation of committees and a few of the CDQ groups have provisions allowing for the addition of other villages to the group, although such villages would still need to qualify for acceptance to the CDQ program under state and federal guidelines.

Phase II: Organization Direction and Plan Development

Organization Direction

The next phase of the CDQ evolution was the determination of organization goals and objectives and the implementation process. The Congressional mandate for the CDQ program that appears in the 1996 Magnuson-Stevens Act reauthorization provides a basic template for looking at activities that CDQ groups have undertaken. This enabling legislation clarified the primary goals of CDQ programs as well as the procedures to be followed to receive the allocation and its benefits. Both federal and state guidelines were established for the plans that CDQ groups were required to prepare to qualify for allocations. The initial program was developed under the direction of the North Pacific Fishery Management Council in the early 1990s with the primary aim to provide communities with the means to develop ongoing commercial fishing activities.

These goals include expansion of employment opportunities in commercial fishing and processing and an increase in capital investment in fishing, processing, and infrastructure. To meet these goals, each CDQ group developed its own approach and tailored its program through local interpretation of needs while following the directives of state program managers.

The setting of group direction required the group to develop long-term goals and more specific project-related goals. This process was conducted primarily by the elected board with input from others in the western Alaska region who were familiar with the fisheries and the economic, social, and cultural characteristics of BSAI villages.

The long-term goals of all the CDQ groups tend to be broad statements such as "stabilize, enhance, and diversify the economy... by participating in the Bering Sea groundfish industry" and "maximize the social and economic benefits... from the harvesting and processing of Bering Sea fisheries." Four of the CDQ groups indicate that a goal for their group is to be a self-sustaining and continuing institution. Three plans address broader economic development objectives beyond commercial fishing. Two groups specify participation in fisheries management institutions to insure the viability of the resources. The specific goals of each group demonstrate significant diversity as might be expected given the resources, history, and cultures of the communities. This variability will be demonstrated in the discussion of Phase III.

Plan Development

To qualify for quota allocations, federal regulations require CDQ groups to submit detailed plans outlining their characteristics, capabilities, and intentions, which require major investments in the acquisition of information, some of it sophisticated and technical. In order to complete these plans, CDQ groups often hire consultants to assist in plan preparation. Some of these individuals have been retained as operational staff. In one case, actual operation of the CDQ group has been contracted to a consulting organization.

Federal CDQ regulations require three types of information: community development information, business information, and a statement of the managing organization's qualifications. The information required under each heading is substantial. Under community development information, topics to be covered include description of projects; allocation requested; project schedule; employment, vocational and educational programs; existing infrastructure; capital uses; short- and long-term benefits; and business information.

In certain of these areas, highly detailed information is required. For employment, the following information must be provided: number of individuals to be employed, the nature of the work provided, the number of employee-hours anticipated per year, and the availability of labor from local communities. Business information includes everything from harvesting methods through product mix to audit control provisions and plans to prevent quota overages. Under managing organization's qualifications are requirements for letters of support from the governing body of each community as well as demonstration of the relationship between the CDQ applicant and the managing organization.

Once completed, the plans are then submitted to the State of Alaska, which reviews the plans for compliance with federal and state guidelines and may return a plan for revision if pertinent information is missing. The state recommends allocation to the governor, who has the option of changing it. Once the state has decided on an allocation, the Council is also consulted.

Although the National Marine Fisheries Service is required to perform a final review of the plans prior to their approval and authorization, this has been basically a pro forma step to date (Ginter, 1997). Given the limited amount of federal oversight, the state process can be controversial because the allocation of quota shares has a multimillion-dollar impact on the CDQ groups. This decision-making process has become highly controversial in the last two cycles of applications and review because of the reallocation of pollock quota from one group to another and the lack of specific reasons given for the shift.

The weighty demands of complying with the planning regulations have resulted in multivolume submissions by the CDQ groups. This, in turn, has translated into a substantial role for consultants and considerable dedication of CDQ group funds to the planning process. These costs, plus additional costs for developing annual plans, maintaining information on harvest levels, exercising fiscal control, and ensuring legal and fiduciary obligations, lead to a rather costly administrative overhead.

An unintended consequence of the plan development and monitoring process has been the location of primary offices of CDQ groups outside western Alaska in Anchorage, Juneau, and Seattle. Only one of the CDQ groups maintains its head office in a community in its region. Most of the fishing companies that partner with CDQ groups in the harvesting of pollock have their headquarters in Seattle, so coordination between CDQ groups is also facilitated by the location of offices outside of the region. An unfortunate consequence of this is that CDQ group administrators become isolated from the daily circumstances in their villages. Over time, residence out of the region may lead to poor communication, cultural dissociation, and may be inconsistent with the goals of the CDQ program.

The planning process has now been through five cycles. Expansion of the CDQ program to BSAI halibut and sablefish in 1995 resulted in awards for halibut to seven organizations. In the fall of 1995, the federally approved three-year extension of the pollock program led to another round of allocations. In September 1997, plans encompassing allocations of additional groundfish stocks and crab were submitted by the CDQ groups and the state made its recommendations to NMFS in early 1998. These allocations were made to the CDQ groups for a three year period from 1998-2000. The harvesting of the crab allocation began in March 1998 and the harvesting of groundfish allocations is scheduled to begin in October 1998.

Phase III: Plan Implementation

Activities undertaken to implement CDQ plans began with the first pollock harvests in the late fall of 1992. However, even before harvests occurred, CDQ groups had a number of things to do. They had to determine the level of staffing necessary to carry out their plans and hire qualified personnel. They then had to develop a process for connecting with fishing firms interested in partnering with them.

Staffing

A fundamental feature of the operation of CDQ groups is that some contractual relationships between CDQ groups and fishing companies are required to implement harvest of the pollock quotas. The reason for this is that the coastal village residents of western Alaska have very little knowledge about the Bering Sea offshore trawl fisheries. In fact, to a certain extent, these fleets have been regarded as the enemy over the last 20 years due to their bycatch of salmon and herring, species that are critical to the subsistence and commercial economies of the villagers. Familiarization with the industry was accomplished by pursuing contacts acquired largely at North Pacific Fishery Management Council meetings or through other segments of the fishing industry with western Alaskan operations.

Proposal Process

A bid process was established to determine a package of benefits that a fishing partner was willing to offer a particular CDQ group. Bid packages were solicited from various firms and the initial discussions were carried out in the fall of 1992. The fundamental component of the proposal was the price a bidder was willing to pay for the right to harvest the pollock. This price could be fixed or could be based on the market price at the open-access fishery. An additional element in the cash component of a proposal might be profit sharing following the actual sale of a finished product. Thus, a complicated matrix of financial choices could confront the CDQ group's board of directors when evaluating proposals with a different mix of options. Additionally, the board had to determine whether to commit to a long-term, multiyear relationship at the outset or whether to proceed on a more conservative basis of one or two years.

While it is conceivable that a proposal could have been made on a strictly cash basis for a group's allocation, several factors militated against such a course of action. In addition to money, CDQ groups were also interested in obtaining jobs for residents, given that unemployment is a serious problem. Thus, packages from bidders including number of jobs, level of pay, duration, and associated amenities (housing, transportation, food, training, and opportunities for advancement) were particularly important to CDQ boards of directors.

Another consideration, not necessarily as important as job opportunities, was the opportunity to obtain equity in vessels or firms, and in several instances, proposals offered CDQ groups these opportunities. These relationships, discussed elsewhere in this report, have had mixed results.

An additional proposal consideration was the amount of training and management opportunities offered that would permit the CDQ group to develop rapidly the capability to manage the full range of their operations. Opportunities for training in other areas of the seafood harvesting and processing industry were considered to be positive aspects of bid proposals. The outcome of the bid proposal process was generally to establish a long-term relationship with a single significant partner rather than dividing the quota among several offers.

Plan Activities

Activities conducted to date by CDQ groups can be grouped into the following "project types" (State of Alaska, 1995, p.17):

- Administration
- General business development
- Employment
- Equity investments
- Processing plant
- Infrastructure
- Access enhancement
- Loan programs
- Fishery development
- Education programs
- Other

A detailed description of the specific investment pursued by the CDQ is provided in Appendix F.

Administration—All CDQ groups have established administrative structures by hiring personnel to conduct the business of the group. There has been a high degree of stability in the executive functioning of most of the CDQ groups, however CBSFA has undergone extensive changes in its executive organization. CDQ groups that have employed persons familiarity with conditions in the villages and the commercial fisheries have been able to function effectively in the new environment.

New administrative activities are underway to manage the multispecies CDQ allocations that were awarded in September 1997, and the fishing season began in October 1998. CDQ groups are required to develop procedures to report target species and bycatch-species harvest levels to the National Marine Fisheries Service and, in essence, to manage their own quota allocations.

It is noteworthy that five of the six CDQ groups have located their main office outside the region they represent. While giving the CDQ groups closer contact with fishing companies and state managers, it has limited their ability to be in touch with and respond to local concerns. Particularly problematic in this regard is effectively providing information to local villages about the CDQ groups and their programs. Many villagers are not aware of the groups or are confused about some of their programs.

General Business Development—Several CDQ groups viewed their mandate to promote local development in a broad fashion. At least two groups proposed to use royalty funds from pollock partnerships to establish businesses that are at best tangentially related to commercial fishing. These programs have not been viewed favorably by state program overseers, who have required businesses to have some significant connection to commercial fishing.

Some examples of general business development conducted by CDQ groups are enhancement of salmon and herring marketing, creation of an Alaska seafood investment fund, establishment of vessel haul-out and storage businesses, and seafood waste conversion. Several of these efforts are in the planning stages and have not yet been implemented.

Employment—All CDQ groups have embarked on programs to provide jobs of various kinds for residents of their communities. All have used their allocation quotas to leverage employment on factory trawlers. While these have been entry level positions, the timing of the work (January to March) has generally fit well with the schedules of villagers from communities where subsistence activities represent an important element of their lifestyle.

A second area of fisheries employment is in shore-based processors of salmon, crab, herring, and halibut. These jobs are generally the result of a cooperative arrangement in which CDQ funds have been invested with an already existing business, or have been used to purchased and revitalize a business. Included in this sector are jobs in ice plants and as dock workers.

Equity Investments—All six of the CDQ groups have made equity investments of one kind or another, and several have set up for-profit subsidiaries.

Fishing vessels have been a common equity investment. Some ventures have been the outright purchase of vessels that are then leased to village residents for participation in local fisheries. Others have been joint venture arrangements with already existing commercial fishing vessels in which the CDQ group purchases a share of the vessel. In most cases, the CDQ groups have purchased less than a majority share of the vessels already operating in the fishery. Partnership arrangements have been established with a variety of different vessels using a range of fishing gear.

One group has taken the process further by building new boats and training village residents in their construction. This has the added benefit of providing new skills in welding and other aspects of boat-building.

Factory trawlers have also been joint-venture equity investments of several CDQ groups. Two of these investments have continued for a number of years while a third, the major investment of one CDQ group, has gone bankrupt, and the CDQ for-profit group established to run the enterprise is presently in an uncertain status. (See Box 3.2).

BOX 3.2   Coastal Villages Fishing Cooperative

The Coastal Villages Fishing Cooperative (CVFC) was established in 1992 as a for-profit corporation for the purposes of participating in the CDQ program. CVFC began to negotiate with Golden Age Fisheries to form a partnership to harvest its pollock allocation. From this partnership, the Imarpiqamiut Partnership (IP) was formed as the manager and owner of the factory trawler, F/V *Brown's Point*. In exchange for 50 percent ownership in the IP, CVFC dedicated all of the quota it was allocated from 1992-1995. The IP made substantial investments in vessel and processing infrastructure; however, the partnership did not prove profitable.

In 1996, the State of Alaska hired a consultant to review the financial status of the *Brown's Point*. Subsequent analysis by the State of Alaska of various financial records indicated that the IP was failing to provide royalties to CVFC under its 1997 pollock "A" season contract, and that the royalties that were supposed to have been collected were significantly below the market rate. Concerns were raised that the processing privileges for the pollock allocation were inappropriately assigned. The CVFC had attempted to permanently convey its pollock processing privileges to the IP, which would have caused a significant long-term loss in revenue to CVFC. The state was further concerned when it learned that the creditors of the IP were anticipating foreclosing on the *Brown's Point*.

In late 1997, the state found that because the *Brown's Point* was undergoing foreclosure proceedings, many of the milestones and goals in CVFC's 1996-1998 Community Development Plan (CDP) would not be met. The state recommended to the National Marine Fisheries Service that unless there were significant changes in the management of the pollock quota, specifically the dissolution of the IP, that the CVFC's 1996-1998 Community Development Plan for pollock should be terminated. The state recommended that the pollock quota allocation be terminated

---

Processing Plants—Shore-based processing plants have been an important activity for three CDQ groups in isolated areas, where a lack of local processing had prevented development of commercial fishing activities. In other cases, older abandoned facilities have been renovated or brought into production. In some cases, investment has improved functioning facility. In at least one instance, an entirely new processing plant was established in a village where processing had never existed. Floating processing facilities have been a third category of equity investment pursued by one group.

As the previous discussion demonstrates, equity investment strategies have

unless the partners and key creditors dissolved the partnership. CVFC appealed the recommendation, stating that it had reached a suitable dissolution agreement with its key creditors to pay off approximately $1.5 million in outstanding debt. In January 1998 the state agreed to rescind its recommendation to terminate the allocation, if CVFC dissolved the IP and reached agreements with its creditors.

Since that time, the Coastal Villages Region Fund (CVRF), a successor to CVFC has been formed to manage the CDQ allocation. The state required CVRF to review and reform its management structure and procedures. CVRF contracted with an independent company to perform the management review. This review provided several recommendations to CVRF and the state has required that these recommendations be implemented prior to the 1999–2000 pollock allocations to be made in September 1998. CVRF has considered the recommendations and has begun to implement them. The CVRF has opened a new office in Anchorage and has hired a financial officer and several managers to oversee the program. The State continues to monitor developments in CVRF closely, specifically communication within the CVRF organization, communication between the state and CVRF, contract negotiation, control over quota, and contract management.

The events in CVFC indicate that the state oversight and the threat of quota termination resulted in changes in the management of CVFC's quota. The mechanism that the state used to create change, recommending the termination of quota allocation unless certain criteria were met, appears to be working. The situation with CVFC points out that regular reviews of the CDQ groups and how they are managing their quota can provide useful information to both the state and CDQ managers. Conceivably, earlier review may have indicated problems in the management of CVFC's quota allocations and could have minimized their financial debt.

varied among the CDQ groups. Some groups have pursued a number of relationships in diversifying their options. One group was initially quite cautious and put off investment until the second funding cycle of the program. At present, they are one of the most active investors and have developed five equity investments in different fisheries.

Infrastructure—The development of fisheries infrastructure has been a major activity of five of the six CDQ groups. One CDQ group, operating in a region with a century-old history of commercial fisheries, has a relatively well-developed infrastructure. Nevertheless, the group has recognized the importance of maintenance and ongoing improvement by establishing an investment fund for future use.

The most frequently undertaken project in the early years of a CDQ program is dock development or improvement. Dredging of harbors has been accomplished in two areas, two communities have received ice delivery systems, while another community has obtained a vessel storage facility. There are additional plans to develop two more buying stations, upgrade a dock, and build a boat ramp.

Infrastructure projects, including roads, storage facilities, and water treatment facilities, have been an especially good way for CDQ groups to demonstrate leadership and creativity by joining forces with regional, state, and federal agencies to leverage CDQ funds and to achieve greater impact.

Access Enhancement—Because of sales to fishermen outside the local area, the Alaskan limited access program for salmon and herring fishery has in several cases lost permits. Lack of access to capital and information about the availability of permits has prevented village residents from purchasing permits. Several CDQ groups have established objectives to assist residents in their areas to gain access to these fisheries.

One group has taken the view that information is the most important missing element and has established a service to link local permit sellers to residents in the region interested in purchasing permits. Another group has established a fund to purchase IFQs in the Alaskan halibut and sablefish fisheries to be fished on the CDQ group's vessels. Two groups intend to establish loan funds to allow residents to purchase either permits or IFQs. Finally, one group is acting as a guarantor to the state for permit purchases through the state loan programs.

Loan Programs—More traditional loan programs for vessels and gear are being run by all six groups. Two have also established small business loan programs for residents wishing to get into fishery support businesses.

Fishery Development—This category includes several activities, each quite different from the other. One CDQ group is examining ways to rehabilitate certain salmon populations, which traditionally have been important to the regional economy. Other groups are initiating fisheries in locations where there has been no commercial fishery. Three groups are investing in exploratory fishing research. Three groups are seeking ways to create new products from their fisheries.

Education Programs—Two basic forms of educational investments in the human capital of village residents have played a prominent role in CDQ activities of all six groups. Educational scholarships, typically about $1,000 per year, have been provided to qualified local residents. These cover any kind of educational program at two- to four-year institutions. The general rationale for this non-fishery activity is that it benefits local communities through improved expertise and earning potential.

The second form of human capital development, for training programs, have focused on skills necessary to participate in jobs on offshore vessels. This training has tended to be in such technical areas such as production line work, refrigeration, machine maintenance, and other jobs related to the technical functioning of the offshore factory trawl fleet. Training for higher level positions on factory trawlers has not yet materialized. One group has proposed development of an observer training program, which could be a good way for village residents to move into the management sector.

Other—One group has established a disaster relief fund to buffer against bad fishing seasons or other disasters that could affect members of their villages.

Phase IV: Performance Review And Revision

CDQ groups have examined their performance and opportunities at several different junctures in their short histories.

A significant component of the performance review is the assessment of how well the group meet the basic guidelines established by the state. This, in turn, is rolled into the periodic cycle of plan development, submission, and defense. These steps have led to plan revision, including elimination or modification of certain programs and activities. Some of these changes have come at the behest of state overseers, while others have been the outgrowth of internal review.

One interesting development has been the interaction among the CDQ groups themselves, me of which have an ongoing exchange of information about guidelines, legal responsibilities, and the state oversight process. In addition, several CDQ groups have initiated small cooperative ventures. More interactive developments are possible, particularly development of forthcoming techniques for managing the new multispecies CDQ allocations.

It is possible, however, that the competitive framework established by the state to reallocate quota share at each renewal cycle could work to preclude cooperation. One way this could be avoided would be to reward cooperation that produces positive results.

4

Evaluation of the Performance of the Community Development Quota Program

The evaluation and performance of the Community Development Quota (CDQ) program requires some standard criteria to measure progress. In general, this can be done by measuring changes in indicators such as unemployment, per capita income, educational opportunities, capital investments, infrastructure, and other quantifiable measures. More difficult is measuring the effects of the CDQ program on the attitudes of the individuals in the community. Factors such as the effects on the local culture, the ability of the program to contribute to self-determination, and the possibility of the program to enhance indigenous uses of modern technology are difficult to quantitatively evaluate. Chapter 2 has described some of the ways in which development is perceived in western Alaska. This chapter attempts to quantify those measures of development that are more easily defined (e.g., employment, income). Additionally, this chapter attempts to evaluate those facets of development that are not easily quantifiable, and attempts to provide means for incorporating them into the consideration of the CDQ program. Understanding if the CDQ program improves the perception of opportunity and provides new avenues and hope for members of the community requires investigation of the social conditions in the communities. To provide a complete evaluation of the effects of the CDQ program, both of these factors are considered in the chapter.

In general, some of the quantifiable factors can be evaluated by comparing conditions before the CDQ program and changes since the program's implementation. However, in some cases the data are not available to adequately measure such changes. Data about the CDQ program that precisely detail the benefits received by the CDQ communities can be difficult to obtain. One of these difficulties is due to the newness of the program and the inability to draw clear conclusions from the limited data that are available. A second difficulty is a State of Alaska law (described on page 92) that certain financial and catch data can be maintained as confidential. These conditions make it difficult to provide a detailed analysis of the benefits received by the CDQ program. However, a review of the data that are available and the nature of the investments of the CDQ groups can be used to evaluate the economic performance of the CDQ program.

The criteria that have been used both by the State of Alaska and National Marine Fisheries Service (NMFS) have tended to rely on quantifiable economic or performance based criteria (see Appendixes D and E). The State of Alaska and NMFS evaluate the CDQ program by measuring criteria such as: the number of community members to be employed and the nature of the work; the number and percentage of low income people in the communities; the number of communities; the relative benefits for the communities and the plans for developing a self-sustaining fisheries economy; and the success or failure in administration of a previous Community Development Plan.

THIS COMMITTEE'S EVALUATION

While this committee kept the State/NMFS criteria in mind during this review, our work was guided most by the charge given to us in the Magnuson-Stevens Act of 1996. Specifically, the committee focused on four broad criteria:

(1) the extent to which such programs have met the objective of providing communities with the means to develop ongoing commercial fishing activities;

(2) the manner and extent to which such programs have resulted in the communities and residents receiving employment opportunities in commercial fishing and processing; and obtaining the capital necessary to invest in commercial fishing, fish processing, and commercial fishing support projects (including infrastructure to support commercial fishing);

(3) the social and economic conditions in the participating communities and the extent to which alternative private sector employment opportunities exist; and

(4) the economic impacts on participants in the affected fisheries, taking into account the condition of the fishery resource, the market, and other relevant factors.

In addition, as the committee became more familiar with the program, it considered additional factors such as the pattern of distribution of benefits, awareness of the CDQ program and its benefits within the community, access by the community to CDQ group directors and managers, and impact of the program on people—whether it increased opportunities, hopes, and self-defined improvements in people's lives. The following sections provide conclusions and recommendations related to: community development strategies, participation and benefits, governance and decisionmaking, environmental and economic sustainability, development of human resources, and program duration.

The North Pacific Fishery Management Council (NPFMC) and NMFS establish the amount of quota allocated to the CDQ program for each of the fisheries based on a decision-making process that incorporates resource assessments, management goals, and political factors going far beyond the CDQ program alone. The amount was not intended to meet all the possible development goals of the communities, but rather to be a contribution, albeit one with significant impact. To date, the amount seems adequate, although over time as the development record builds this may need to be monitored.

COMMUNITY DEVELOPMENT STRATEGIES

The strategies pursued by the six CDQ entities have varied considerably. A detailed description of the investments pursued by the CDQ groups is provided in Appendix F. The range of activities reflects different resource situations and economic circumstances among the groups, as well as different goals. The perception of a tenuous future for the CDQ program (in a political sense) has put the governing boards in a difficult position. The fear that the program may be terminated has induced some groups to seek investments that will yield economic returns as quickly as possible. In a few instances, this strategy has resulted in unwise investments. Ironically, where this strategy has resulted in investments that have turned out well the boards are vulnerable to criticism that they have pursued monetary gains while ignoring the simpler needs and aspirations of their members in the villages. Indeed, the committee found that some groups had pursued a strategy to maximize economic returns, while others paid close attention to the participation of members of their villages, often at a less-than-maximum financial return to the CDQ corporation.

Another form of uncertainty has equally perverse influence on the behavior of CDQ boards. While some may be moderately confident that the CDQ program will persist over time, they fear that their particular allocation of a share of the total CDQ quota is highly uncertain. This uncertainty arises from a sense that the criteria used by the State of Alaska to allocate individual shares of the total quota are unclear. There is a concern that if a group is perceived by the State as receiving "too much" income their share of the total allocation may be reduced and given to another group with greater needs. Conversely, if a group is not performing well it may lose its share of the total allocation in the future. The committee finds these various forms of uncertainty to have undesirable effects on the development strategies chosen by the various boards of directors.

There is another aspect of the CDQ program that warrants comment. The

This trawler has just delivered CDQ pollock to a shoreside plant in Akutan. Involvement in commercial fisheries requires significant economic investment. To some extent, the plans and implemented activities of the CDQ groups were influenced by uncertainties about the duration of the program and the stability of the shares allocated, which caused some groups to pursue strategies designed to maximize economic returns quickly while others focused on longer-term investments. (Photo by Craig Severance.)

NPFMC, the NMFS, and the State of Alaska have determined that all investments made by the CDQ groups must be in "fishery related" activities. This restriction means that although the CDQ program has two objectives—community development and fishery development—"community development" is defined as "fishery development." The committee finds this strict requirement to be of dubious merit. There are, to be sure, advantages to a fisheries program that encourages continued investment in, and improvement of, fishery resources and fishing capacity. To the extent that there are viable fishery-related investments in the coastal villages that promise reasonable returns on investment, they should be pursued. However, we can foresee a time when this restriction on investment opportunities will force the CDQ boards to make investments that may not promote economic diversity and sustainability at the village level. It is also possible that the available sound fisheries investments in many villages will ultimately be exploited, in which case the restriction will force some CDQ boards to undertake less than ideal investments.

A more compelling argument is that "community development" should be seen as broader than just fisheries development. While all CDQ villages are, by law, within 50 miles of the Bering Sea, some of the subsistence economic activity in these villages is land-based. In addition, there are many investment needs in these villages that would contribute materially to "community development." Among the alternatives are development of general infrastructure, health clinics, recreation centers, schools, improved roads, water and sewerage systems, and fire protection.

Conclusion

While the Community Development Plans prepared by each of the CDQ groups are similar in some important respects, the specific elements included vary considerably among groups. Each CDQ group pursues income from the large scale pollock fishery through royalties and employment, and each seeks to develop nearshore fisheries using smaller vessels. The diversity of infrastructural investments, training programs, and financial strategies adopted by the CDQ groups does, in our judgment, appropriately reflect varying circumstances and reasoned approaches to diverse problems. To some extent the development plans were shaped by uncertainty about the duration of the CDQ program and by the restriction that the CDQ plans must focus on fishery development. For example, the uncertainty may have encouraged at least one CDQ group to seek a quick financial gain by convey-ing their quota rights in perpetuity. We found this permanent conveyance to be inconsistent with the philosophy and intent of the CDQ program. Finally, the economic and cultural development of these communities may at times be advanced through non-fishery employment or investments. Hence, we found no strong reason to require the communities to use funds generated from their CDQs to invest only in fisheries.

Recommendations

• We recommend that the State of Alaska prohibit permanent conveyance of community development quotas into the hands of commercial enterprises out-side the communities. An important aspect of the community development sought in western Alaska is the continuing and direct involvement of local people in fisheries of the Bering Sea. Sale of the CDQs to outside interests will create an inappropriate separation of the people from the regional resources.

• We recommend that the restriction that CDQ revenues to be invested only in fishery-related activities should be removed, at least for some portion of the revenues. Many of the communities will find that fishery investments are still the ones they wish to undertake. However, since community development is broader than fishery development, funds should also be available for other activities that will enhance community infrastructure or land-based economic activity. This broadening of the allowed investments would also remove uncertainty about whether particular investments are indeed "fishery related" and thus allowable under current rules.

PARTICIPATION AND BENEFITS

The CDQ program has had a positive economic impact on western Alaskan communities. During the first four years of operation, the six CDQ groups collected over $92 million in gross revenues from fishing partners (Table 4.1). This revenue is derived primarily from royalties received from offshore pollock fishing partners. Somewhat more than half of the total revenue has been accumulated for use in future fisheries development projects. The other half has been spent on project administration, training, fishing equipment, infrastructure development, and commercial subsidiaries of the CDQ organizations. A detailed description of the investments pursued by each of the CDQ groups is available in Appendix F. Current administrative expenses, in the 20 percent range, are high and should be expected to be trimmed as the groups mature. During 1992-1996, a total of $3.8 million was spent on a variety of training projects involving 3,443 people, and western Alaska residents were employed on a total of 2,503 fishing trips that range from a few days to several weeks in length (Table 4.1).

The CDQ program has enhanced the employment of western Alaskans in the commercial fishing industry—before the program, employment on factory trawlers and in on-shore processing plants was not generally available to the people of the CDQ villages. The categories in which employment increased are: (a) the

TABLE 4.1 Total Revenues, Expenses, and Income for Six CDQ Organizations

	1992-1996 (million $)	1996 (million $)
Total Revenues	$92.71	$24.24
Project Expenses	$25.86	$6.93
Administrative Expenses	$17.22	$5.63
Tax	$0.04	
Net Income	$50.68	$11.94
Training Expenses	$3.80	$1.22
Wages Earned	$19.99	$6.60
Fishing - Positions/Trips	2503	1101
# Training Opportunities	3443	1126

SOURCE: DCRA, 1997

pollock fishery, which is separated into the winter fishery for roe-bearing pollock, called the 'A' season, and the autumn fishery which is called the 'B' season, (b) "other fishing" which includes fishing for species other than pollock or traditional salmon and herring fisheries, (c) "other positions" which includes onshore fish processing and other jobs not with the CDQ organizations themselves, and (d) internships with the pollock fishing partners (Table 4.2). Over the four years, the average income earned per position was $6,567 in the pollock 'A' season, $5,042 per position in the pollock 'B' season, $3,303 per position in "other fishing," and $4,917 in "other positions" (Table 4.2). The data do not indicate how frequently individuals participate in more than one of these categories. It is possible, for instance, that a resident active in both pollock seasons and in other fishing could earn cash income on the order of $14,900. The significance of this income supplement can be judged by comparison to annual cash incomes per household, which averaged $30,180 for all western Alaska native villages in 1990, but ranged from a low of $11,340 in Savoonga to $54,070 in King Salmon in 1990.

As shown in Table 4.3, the distribution of employment among these categories varies widely among the CDQ organizations. For example, a relatively large proportion of new fishing employment in BBEDC's villages occurred in the offshore pollock fishery, while at the other extreme CBSFA placed almost all of their new positions in other local fishing. The other four organizations spread the employment among pollock, other fishing, and other positions with an particular emphasis on "other fishing." Employment in the "other fishing" sector of the fishing industry may change with the implementation of the multispecies CDQ program in 1998. Scarce information about the amount and types of fishing occurring under the "other fishing" category makes it impossible to characterize

TABLE 4.2 Employment Attributed to CDQ Organizations in 1996.

	Employment	Wages	Income per Position
Management/Administration	103	$1,843,000	$17,900
Fishing			
CDQ Pollock 'A' Season	161	$1,057,000	$6,570
CDQ Pollock 'B' Season	136	$686,000	$5,040
Other Fishing	691	$2,283,000	$3,300
Other Employment	106	$521,000	$4,920
Internships	32	$202,000	$6,330
Total	1229	$6,593,000	$5,360

SOURCE: DCRA, 1997.

TABLE 4.3 Indicators of Economic Impact of CDQ program in Western Alaska Villages in 1996. "Positions" represent various employment levels, ranging from a few weeks of pollock fishing to full time staff positions.

Year 1996	APICDA		BBEDC		CBSFA	
	Positions	Wages	Positions	Wages	Positions	Wages
Management & Administration	15	$257,600	6	$241,800	17	$259,120
Pollock 'A' Season	4	$30,600	57	$217,200	7	$29,150
Pollock 'B' Season	10	$72,200	43	$141,600	0	$0
Other Fishing	55	$605,100	42	$128,100	146	$454,000
Other Positions	72	$308,700	0	$0	9	$39,670
Internships	0	0	22	$63,900	1	$13,620
TOTAL	156	$1,274,200	165	$978,100	180	$795,860

TABLE 4.3 (part 2)

Year 1996	CVFC		NSEDC		YDFDA		TOTAL	
	Positions	Wages	Positions	Wages	Positions	Wages	Positions	Wages
Management & Administration	7	$248,870	22	$546,740	14	$142,240	103	$1,843,370
Pollock 'A' Season	52	$167,040	45	$216,100	32	$358,000	161	$1,057,300
Pollock 'B' Season	36	$108,710	61	$248,710	7	$114,570	136	$685,780
Other Fishing	179	$229,430	183	$336,980	86	$528,900	691	$2,282,510
Other Positions	1	$18,600	11	$11,610	13	$142,360	106	$521,240
Internships	3	$36,710	1	$20,000	5	$68,250	32	$202,470
TOTAL	278	$809,350	293	$1,380,140	157	$1,355,020	1229	$6,592,670

SOURCE: DCRA, 1997.

One of the goals of the CDQ program is to increase employment for people from rural areas of western Alaska where jobs are scarce, which then brings benefits to their communities. Before the program, employment of western Alaskans in the commercial fishing industry was limited, but opportunities on both factory trawlers and in on-shore processing plans are now showing increases. (Photo by Craig Severance.)

this activity in more detail. This is due in part to State of Alaska laws regarding the confidentiality of financial data.

Because data are not available to indicate who in the community is receiving employment, it is difficult to determine whether the disadvantaged members of the community are receiving employment in the CDQ program. It is possible that those members of the community who have experience with other organizations and corporations, such as Alaska Native Claims Settlement Act (ANCSA) corporations, could be deriving greater benefits from the CDQ program. More data would be needed to assess the distribution of employment benefits within the particular communities.

According to testimony received by the committee and the information gathered during the site visits undertaken by committee members, several factors about the participation in the CDQ program are apparent. The main participants from the villages come from three groups: (1) the community and regional leaders who manage the development projects; (2) the mobile young adults who are able to work on fishing vessels or accept university fellowships or training programs outside the community; and (3) adults with families who participate in community-based projects such as fish processing plants and local, small-boat fisheries for halibut, sablefish or crab. The first group involves relatively few people, but is of strategic importance because it includes people becoming more capable of business management through familiarity with corporate structures and procedures. At present, the second two groups are larger and constitute the target populations for the community development program. Increased cash incomes in these two groups entail definite development benefits—an amelioration of major social problems and the achievement of a measure of self-determination, as well as the alleviation of poverty.

The participation of young women and men in the wage and educational programs is especially salutary, given that this group has been at greatest risk of the suicide, alcoholism, and drug abuse problems that have plagued western Alaska native communities (NRC, 1996). The strict anti-drug conditions of trawler employment have had beneficial effects, according to local testimony, and are much appreciated by community elders. At the same time, many of the young people have made productive use of their wages, such as investing in snow machines, all–terrain vehicles, and other equipment needed for subsistence activities for themselves or their families. Moreover, it can be foreseen that the educational and training programs will be at least as beneficial locally as the fisheries employment—in ways not necessarily imagined by earlier notions of "community development." To appreciate this difference we need to explore the concept of "community" and its future in western Alaska.

Overall, testimony to this committee indicated that the program has contributed to improved understanding of business administration, corporate structure and procedures, and technical skills of village residents. This practical training facilitates long-term enhancement of the cash earnings potential of western Alaska villages.

Western Alaskans have become involved in a fishery (pollock) that previously seemed beyond their capabilities. Moreover, the CDQ groups in Yukon Delta, Norton Sound, and Bristol Bay have shown a capacity to link up with municipal agencies to co-sponsor important infrastructure projects that would otherwise have been left undone due to a lack of initiative and funds. The management of the corporate structures and boards responsible for strategic and operational decisions has improved from the earliest efforts at CDQ programs in 1992. While at least two CDQ groups experienced serious shortfalls in business planning and investment strategy, these appear to have been addressed through interaction between State oversight authorities and the CDQ group leadership. Continued enhancement of these management skills will follow further experience, formal training, and State oversight.

Improvement of program management capabilities will accelerate the development of a cadre of western Alaskans who are capable of mediating between the rural subsistence culture and the urban westernized cultures. These people are a critical resource for the development of western Alaskan communities, which relies upon local subsistence resources, commercial fishing activities, and external wage-earning opportunities.

The training, education, and employment opportunities received from the CDQ program may assist even those village residents that chose to move to larger communities on either a long-term or short-term basis, because often their transition is hampered by lack of skills and employment opportunities. Although training and education may encourage some residents to leave at some loss of human capital, even this can have benefits because those who leave tend to maintain contact, sometimes building skills further and then returning and in any case providing increased links to the rest of the state.

Through testimony and direct observation, the committee found another important value of the CDQ program, which is less tangible but in some respects the most promising for development and the most appreciated by the people concerned. The general heading of these values would be something like "self-determination"—a condition of autonomy which both the larger American society and traditional native custom take to be a supreme virtue. Everything about the subsistence life style encourages and rewards self-reliance. The depth of the despair that ensues when such autonomy is frustrated—paradoxically, under modern conditions, by the lack of cash for traditional activities—is a measure of the importance of self-reliance for the native communities. Another indicator is the enthusiasm for the CDQ program voiced in several villages visited by the Committee. However, it must be noted that awareness of the CDQ program by residents of some of the participating villages was inconsistent, and in some places quite low (Hensel, 1997). These findings underscore the necessity of appropriate communication between the CDQ groups' administrators and member villagers.

In various ways the CDQ program is giving western Alaskans a new measure of control over their lives. Training in such activities as welding and boat-building brings these services into the village and under local management. Similarly, the wage opportunities (e.g., on factory trawlers) not only connects the village to the larger economy, but provides a means to control that connection. Individuals choose when, and under what conditions, to participate in the commercial fisheries of the BSAI. To the local people, the CDQ program benefits bring valued independence and greater self-determination.

Conclusion

The CDQ program has had important impacts on western Alaska communities. The CDQ program—through its access to a small share of the fishery resource in the Bering Sea—represents a reallocation of economic opportunity to residents of western Alaska. This new economic opportunity

consists of two distinct dimensions. The first dimension concerns the opportunity (and the need) for western Alaskans to create new governance structures and processes around the management and harvest of this newly available access to fishery resources. The mere fact of this opportunity to organize and mobilize residents of western Alaskan villages around the fishery resources of the region represents one important element in the process of development. These new structures and processes have, in addition to their concern for fishery development, begun to play an additional role in the communities. For instance, village residents with questions about social security checks, other employment opportunities, or college programs come to the CDQ offices for advice. The second dimension concerns the more traditional "outcome" oriented impacts. Significant revenues have been generated to support fishery-related investments in the villages. Employment opportunities have expanded for young people able to take advantage of relatively well-paying jobs on factory trawlers and in processing plants. Finally, general education and training programs have brought benefits to other members of the villages in western Alaska.

Recommendations

• The Community Development Plans should be careful to balance the mix of local fishing opportunities with wage-earning opportunities through fishing partners. This is important because local fishery development can benefit less mobile village residents, while wage-earning opportunities in the industrial fleet are especially important for younger, more mobile adults. A focus on local fisheries opportunities for permanent village residents, where they exist, will help tie the CDQ program to the village economies.

• To improve the effectiveness of developing a well trained workforce, the CDQ groups need a strategic plan for education and training programs. This would include internships and technical training for direct employment with the industrial fishing partners of the CDQ groups, formal university education in fields pertinent to the development goals of native residents, and training of administrators and board members of CDQ organizations. The ultimate objectives would be to develop both the business acumen and labor productivity of village residents.

GOVERNANCE AND DECISIONMAKING

The overall structure of the CDQ program established by the North Pacific Fishery Management Council was reaffirmed and codified in federal legislation during the reauthorization of the Magnuson-Stevens Fishery Conservation and Management Act of 1996. Decisions regarding management of various aspects of the Bering Sea fishery and the CDQ portion of that fishery are made at a many levels, including federal, state, and local. The six CDQ groups operate in a system of governance which reviews their decisions in several different ways. Generally, regulatory power over fisheries resides at the federal level, with the National Marine Fisheries Service providing enforcement of fishing regulations. Business management decisions are reviewed at the state level. The North Pacific Fishery Management Council makes broad policy decisions about the structure of the fishery. The CDQ groups make decisions about how to harvest their quota and select partners to assist them in their harvest. Halibut is regulated by the International Pacific Halibut Commission with input from the North Pacific Fishery Management Council.

The structure of the CDQ portion of the system was influenced by Alaska's experience with village and regional corporations created by the ANCSA (see Box 4.1). In its structure, ANCSA created both for-profit and nonprofit corporations, and some of these corporations experienced severe business difficulties. The system of oversight designed for the CDQ program was motivated, in part, by a desire to avoid the problems that developed with the ANCSA corporations.

The following text provides an overview of national, state, and local decisionmaking structures.

Federal Oversight

Federal oversight of the CDQ program involves several elements. The North Pacific Fishery Management Council allocates a portion of each species to the CDQ program as a whole, while the State of Alaska allocates the overall Community Development Quota to the specific CDQ groups representing the eligible communities. The National Marine Fisheries Service has codified the general structure of the CDQ program in 50 CFR 679.30-34. These regulations specify the eligibility criteria for communities to participate in the CDQ program, and the specific content of the Community Development Plans. An additional qualifying criterion is that at least 75 percent of the board of directors of the applicant organization be resident fishermen of the community or group of communities that are applying.

The Community Development Plans need to contain information on three topics:

(1) Community development information, such as employment, vocational and educational programs, existing fishery-related infrastructure, sources of new capital, and a schedule for transition to self-sufficiency in fisheries rather than dependence on the CDQ program.

(2) Business information, such as the method of harvest, description of business relationships (including how proceeds are divided among the parties), a general budget, a list of capital, a pro forma cash flow and break-even analysis, and a projected balance sheet and income statement.

BOX 4.1 The Alaska Native Claims Settlement Act

The Alaska Native Claims Settlement Act (ANCSA) was signed into law on December 18, 1971 (43 U.S.C. 1601 et seq.). The Act was the result of an extensive and protracted effort by Alaska Natives to receive compensation for land held prior to the arrival of Euroamericans in 1741 and subsequently lost. The Act was influenced by various events in the disposition of land in Alaska including the sale of Alaska by Russia to the United States in 1867 (The Treaty of Cession), the Organic Act of 1884 that established protection for Alaska Native land rights, and the Statehood Act in 1959.

Alaska Native concerns about land status increased following statehood as the state began selecting lands in the vicinity of native villages. In 1966, a statewide native organization was created, the Alaska Federation of Natives, to pursue Alaska Native claims issues. Through its efforts, then-Secretary of the Interior Seaton imposed a freeze on further state land selections until native land claims were settled. Pressure for a settlement increased dramatically when land rights were needed to build a pipeline to transport oil discovered at Prudhoe Bay in 1968. The combination of native efforts, the land freeze, and desires for oil development eventually led to the passage of ANCSA (Arnold, 1976).

ANCSA provided $962,500,000 and 44,000,000 acres of land to Alaska Natives as settlement of their aboriginal claims to land in the state. The land and funds were not transferred directly to Alaska Native governmental entities, however. Instead, ANCSA created 12 regional for-profit corporations and over 200 village for-profit corporations. These corporations received title to the conveyed land and settlement moneys. Alaska Natives became owners of the corporations as shareholders based on where they resided when ANCSA was passed. Each Alaska Native received 100 shares in the corporations for which he or she qualified. The regional corporations hold title to subsurface rights to their lands as well as those of the village corporations within their region.

At first, the opportunities provided by these corporations appeared promising and Alaska Natives were optimistic that substantial economic benefits would follow development activities. But problems emerged as the implementation of ANCSA proceeded. In many cases, the corporate structures created by ANCSA did not provide the hoped-for gains. Lack of familiarity with for-profit organizations, limited managerial expertise, and confusion led to circumstances that allowed advisors and consultants to profit without providing real opportunities to the native corporations. Neither oversight nor advisory functions were provided legislatively by federal or state agencies to assist with these problems. In many cases, particularly in western Alaska, the lands selected by the village corporations were chosen primarily based on subsistence needs and

continued

BOX 4.1 Continued

offered few or no prospects for development or job creation; as a result, many of the corporations simply ceased to function. The lands they selected were primarily based on subsistence needs. To some extent, the cash dividends received simply flowed from the shareholders in the communities to industries and services outside the region for imported goods and services. The fortunes of the regional corporations were directed in large part by the development potential of the lands they were allowed to select. Because land in western Alaska had little development potential, corporations in what is now the CDQ region had a difficult time. Bering Straits Regional Corporation was in bankruptcy in the late 1970s. Calista, the regional corporation of the Yukon-Kuskokwim region, was assisted through Congressional legislation in 1988, paying the corporation a substantial bonus over the appraised value for lands turned over to federal refuges in the region. The problem of corporate insolvency and potential loss of lands has been a major issue as Alaska Natives have sought greater economic and political self-determination during the last decade (Langdon, 1987). Corporate solvency was addressed by Congress in the mid-1980s by creation of a program that allowed some Alaska Native corporations to partially refinance.

Some ANCSA corporations have been more successful than others at defining goals, creating business and employment opportunities, and identifying funds for infrastructure development. They are also called on to meet social goals — education programs and scholarship funds — and to be political advocates for village residents. Although the experience with ANCSA has been mixed, overall it is clear that some important components of the more successful programs include the development potential of conveyed lands; well-trained, responsive managers; clearly defined goals; effective community communications and participation; and oversight by village and regional administrators acting in the best interests of their constituents.

SOURCES: Alaska Natives Commission, 1994; Arnold, R. 1976; Langdon, 1987; Case, 1984; Berger, 1985.

(3) Statement of the managing organization's qualifications, including a description of the management structure, resumes of key personnel, and a list of "management qualifications" to provide evidence that the organization can prevent quota overages. Additional information such as letters of support from the communities, demonstration of management and technical expertise, including the balance sheet and income statement for the past twelve months, should be provided as well.

The NMFS regulations set out five "factors" that appear to be criteria NMFS requires the State of Alaska to apply when judging the completeness and adequacy of the Community Development Plan:

1. The number of community members to be employed and the nature of the work,

2. The number and percentage of low income people in the communities, connected to the opportunities to be provided to them,

3. The number of communities,

4. The relative benefits for the communities and the plans for developing a self-sustaining fisheries economy, and

5. The success or failure in administration of a previous Community Development Plan.

After the applications and allocations of quota are approved, the regulations require annual reports to be submitted to NMFS. These reports must include an annual progress report, and an annual budget report, reconciling projected and actual income and expenditures.

Substantial amendments to a Community Development Plan must be approved by the State of Alaska and the NMFS. A procedure has been put in place to expedite this stage of review. After the State reviews the amendment, the amendment is approved if NMFS does not disapprove within a 30-day time period. Once both the State and NMFS have approved, the amendment is accepted. Major capital expenditures are defined as substantial amendments. Since the Community Development Plans are completed each three years, changes to capital expenditures are likely and the review of such amendments is an important part of oversight by the State and the NMFS.

State Of Alaska Oversight

Based upon its evaluation of the Community Development Plans submitted by the six CDQ groups, the State of Alaska allocates a percentage of each species' CDQ to the communities. Evaluation of the Community Development Plans is coordinated by the Department of Community and Regional Affairs, with participation by staff of the Department of Commerce and Economic Development and the Department of Fish and Game in a CDQ evaluation group. The CDQ team reviews the Community Development Plan applications and, after a public hearing to resolve questions, provides summary comparative information to a committee made up of the commissioners of each of the three departments just listed. The three commissioners together make a recommendation to the Governor, who makes the final decision.

In allocating the quotas, the State faces a complicated multicriterion decision-making problem. The need for the fair application of a set of criteria to the applications was raised at our hearings and in our site visits. Multi-criteria decision-making is difficult. As Arrow and Raynaud (1986) have shown for the general case, such decision-making is troubled by two tendencies: either one criterion appears to take over as the single criterion, or decisions appear to be inconsistent. These dangers exist whether or not the analyst applies a "scoring system," which implies a single unit of measure for all of the criteria. Use of a scoring system tends to increase the probability of being inconsistent. Thus, it should be no surprise that the people most affected by the application of the many criteria that the State uses find the outcome somewhat difficult to fathom. The problem of allocating resources based on many criteria is intrinsically a difficult problem. In recent years, multi-attribute decision analysis has improved, but it remains complex.

The application form for the multispecies CDQ allocation in 1997 lists 16 criteria for use by the State (see Box 4.2).

The 16 criteria used create a difficult ranking problem. A reorganization and

---

BOX 4.2
Criteria Used by the State of Alaska to Evaluate the
1997 CDQ Allocation

1. The application's objectives.
2. Realistic measurable milestones for determining progress.
3. Previous ability to manage a Community Development Plan.
4. Methods for developing a self-sustaining local fisheries economy.
5. Level of career track employment and training opportunities.
6. Capital or equity generated for local fisheries investment.
7. Profit-sharing arrangements.
8. Diversity in harvesting/processing partners and modes of operation.
9. Coordinated activities with other CDQ groups(s).
10. Investments with experienced industry partners.
11. Ability of a CDQ group to maintain control over allocations.
12. Involvement and diversity in all facets of harvesting and processing operations.
13. Depth of seafood related infrastructure development.
14. Stimulation of Alaska's economy in both CDQ and non-CDQ communities.
15. Conservative and sound management principles in the fishing plan which provide for full retention and utilization of quota.
16. The development of innovative products and processing techniques aimed at conservation and maximum utilization.

simplification of the criteria was provided in the State's recommendation to the North Pacific Fishery Management Council through a letter from the commissioner of the Community and Regional Affairs to the North Pacific Fishery Management Council dated September 23, 1997 (see Box 4.3).

The order of the criteria proposed in the letter is different from that provided in the application booklet. For instance, the first and second criteria in the state's letter, population and income, are not mentioned in the application. The first criterion in the application booklet, the application's objectives, are not listed in the letter. The criteria given in the letter are also different from the ones listed at 50 CFR 679.(d)(5)(iv). For instance, the goal of developing a self-sustaining fisheries economy listed in the CFR is present only by implication in criteria 3, 6, and 7.

Given the difficulty of allocating quota based upon these many criteria, discussion of the overall purpose of state oversight is important. Is the overriding goal of state oversight to provide an equitable division of the quota allocated to the communities? Or is it to provide a check upon possible mismanagement (e.g., poor investments, misallocation of royalty payments) by the community development management organizations? It appears that the system is working as a mix of these two goals. The actual allocation of quota seems to be driven by two criteria: population and income levels. The CDQ regions with the most people or the most poverty or both tend to get a larger quota. But the allocation proce-

---

BOX 4.3
Simplified Criteria to Evaluate CDQ Allocation
Proposed by the State of Alaska

1. The number and population of eligible communities represented in each group.
2. The level of income, unemployment and other indicators of social and economic well being that demonstrate the need for the allocation.
3. The merits of the proposed investment, employment, education and training programs.
4. The qualifications of the management organization to effectively manage the quota.
5. The contractual relationships between the applicant and the harvesting and processing partners.
6. The degree of success as measured by performance of the management organization.
7. The ability of the Board of Directors to effectively maximize the benefits of the program to the region.

dure can be used to serve the purpose of minimizing potential mismanagement by application of criteria 3 to 7 of the summary criteria listed in the State's letter to the North Pacific Council.

In a desire to prevent mismanagement by CDQ groups, the State of Alaska is trying to use its oversight powers to induce better performance. When this Committee visited Alaska, it held an information-gathering workshop and spoke with many of the leaders of the groups; the efficacy of the State's oversight was a major source of discussion. In the first reallocation of the pollock CDQ, St. Paul had its allocation reduced and the Yukon Delta had its allocation increased; part of the reason for the changes appear to have been response to management performance. At issue in 1997 was the review of the Coastal Villages Fishing Cooperative (CVFC) and its financial problems (described in Chapter 3). But there was a concern expressed that long-range planning is difficult when the whole amount of quota is at risk during reevaluation.

In its September 23, 1997 letter to the Secretary of Commerce, the State of Alaska also recommended that the CDQ group's multispecies allocation:

> be held in abeyance until the following conditions are met: 1) submission of plan amendment dissolving the Imarpiqamiut Partnership between CVFC and Golden Age Fisheries; 2) steps be taken to strengthen its management structure and (3) an agreement to perform an open RFP process for the lease of the pollock allocation in 1999 and beyond.

The State was willing to exercise its leverage in allocation of quota in order to make one of the CDQ groups improve its performance. It began by using the multispecies allocation to induce changes. But if that was not enough, the letter continued, "The state also intends to recommend termination of the 1998 pollock quota to Coastal Village Fishing Cooperative unless the IP partnership is dissolved."

By the first and second criteria recommended in the September 23 letter, the large and poor region represented by the Coastal Villages Fishing Cooperative would merit a large allocation, rather than take away its allocation entirely. However, the State decided to threaten action in order to obtain compliance from the CDQ managers in the Coastal Villages region. On October 31, 1997, the executive director of the Coastal Villages Fishing Cooperative indicated that his board of directors had decided to dissolve the partnership. The Coastal Villages region has since begun proceedings to dissolve the Imarpiqamiut partnership, formed a new organization to oversee the management of the CDQ (the Coastal Villages Region Fund) instituted a management review project, and changed harvesting partners. In a January 28, 1998 letter from the deputy commissioner of the Department of Community and Rural Affairs to NMFS, the State indicated that based on these changes, it was rescinding its recommendation to withhold quota from CVFC. The State's letter was unprecedented, but because the threat seems to be pushing the group to make significant changes, there is no test yet of what will happen in the future if the State needs to enforce a similar threat in some other situation.

One issue in this type of oversight is the threshold level which induces a reallocation of quota from one CDQ group to another. In evaluating the return on investment, it is not clear how broadly "returns" are identified or counted. For instance, education is valued for its relevance but the increase in wages that an educated community member receives over that of others is equally important. The return to the community from having a more skilled worker may be more economically valuable than the increase in that worker's wages, particularly in a poor community. Enhanced skills and education can have a broader influence on community capabilities and long-term economic prospects although these effects may not be fully reflected in individual incomes, especially if the skills provided by education are diverse and act in concert with each other. Investments in infrastructure pose a similar problem, in that the increased productivity of the community may not show up on the books of the CDQ organization. In order to adequately assess the impact of improved education and infrastructure, a much more detailed study would need to be undertaken.

The criteria of both the NMFS and the State of Alaska appear to stress observable impacts which occur in the financial and narrative reports that the CDQ groups submit quarterly and annually. To what extent can the communities define words such as "the merits of the proposed investment, employment and training programs," or "effectively maximize the benefits of the program to the region"? If these terms are defined in terms of market-oriented economic evaluation, the State may overlook the contributions to the modern subsistence economy. The first criterion listed in the Community Development Plan application for multispecies allocation was "the applicant's objectives." This criterion, however, is not present in the letter to the North Pacific Fishery Management Council, and it is not present in the federal regulations. This would suggest that the State means to evaluate the applicant's objectives according to other criteria, rather than to allow the applicant to explain the objectives of its plan and to have the plan evaluated based upon those objectives. The application materials, however, can be interpreted to mean that an applicant can have some influence on the definition of success for its own community development program. If success is defined only in traditional market terms, some of the communities' definitions of success may not be fully recognized in the oversight process.

CDQ Groups

The CDQ groups are organized as corporations under Alaskan law; five of the six groups were organized as nonprofit corporations initially, and now the sixth is also a nonprofit corporation. In all cases, the members of the corporation are the communities served by the CDQ organization, not individuals from the communities. Thus, boards of directors are selected in municipal elections or selected by municipal governments, not at meetings of the shareholders of the corporation as would be the case with a corporation whose owners or members are individuals.

The selection of such an organizational form has implications for accountability to the communities. Corporations with individual memberships may face problems in that the many individual members, because of the difficulty of communication and organization among themselves, are unable to exercise major control over the board of directors and the management of the organization. Representatives elected by communities or selected by community governments would be fewer in number and more able to communicate among themselves in selecting the policies of the corporation. But the members of the communities would have representation only through their elected representatives. The accountability question is thus moved, in part, back to the system of local elections within the villages. Village residents do not have to learn a new system of governance, that of the CDQ group. Rather, residents use their existing system of governance to select their representatives to the CDQ corporation.

With one major exception, the Committee heard few complaints about this governing structure, and some major support for it. The fit with local governing approaches seems to have been fairly good. Some constituents from some areas indicated that they preferred an individual membership approach rather than a community membership approach but have not been able to change the current approach. In some of the communities the committee visited, there were isolated concerns about aspects of the governing structure, but the committee did not hear widespread criticism about the governance structure as a whole.

One notable characteristic of Alaska's laws regarding corporations is that financial data and catch data can be defined as confidential. At 6 AAC 93.070, the regulations state:

(c) Good cause to classify a record as confidential under this section includes a showing that
(1) disclosure of the record to the public might competitively or financially disadvantage or harm the eligible community, qualified applicant, or managing organization with the confidentiality interest, or might reveal a trade secret or proprietary business interest; and
(2) the need for confidentiality outweighs the public interest in disclosure.

The result is that the management and board of directors of a CDQ group can, if they wish, keep such data away from the public. This is a characteristic of Alaskan law, not a necessary characteristic of the CDQ program itself. The committee observed some dissatisfaction with the communication of information from the management of the CDQ groups to their constituents.

Although the Community Development Plan that is submitted to the State is a public document—with the exception of financial data, which is in an appendix—

it may not be widely available and accessible to those in communities who wish to read it. The state's regulation, at 6 AAC 93.070, states that non-confidential information "submitted under this chapter by a Community Development Plan applicant and in the possession of the governor or governor's designees . . . are open to inspection by the public during regular office hours." A trip to Juneau would be required to inspect the records, and no authority to make copies is provided.

Given the extensive amount of review that the State and the Federal Government exercise over the CDQ program, one would expect that management of the CDQ groups places considerable effort in compliance with that oversight. Information would be readily available to the State of Alaska, less so to community members. A major issue is the extent to which communication with the State potentially reduces the responsiveness of the management of the CDQ groups to the desires of their constituents in the communities they serve.

One part of this issue is the location of the main office of the CDQ group. Only one of the six headquarters is located in the region it serves. The other headquarters are located in Anchorage (2), Juneau (2), and Seattle (1). Several of the groups have subsidiary offices in one or more of their communities. Why do so many have their main office outside of their communities? Two answers have been provided: to improve communication with the State, and to improve business communication, primarily with the industry partner. For instance, the process of requesting bids and selecting a business partner is easier to carry out from one of the major cities. The governance structure also orients management toward the regulators; quarterly and annual reports are due to them and the confidential financial data are available only to the regulators. There are no regulations providing for regular reporting to constituents in the communities. As a result, managers of CDQ groups might quite understandably tend to orient their work toward those who receive regular reports and have the power to change allocations rather than to the constituents in the communities. The balance of communication is thus shifted toward the State managers and away from the communities they serve. The idea of "community development" would seem to require that the people in the communities should have more influence over management than others do. For this reason, reporting to the communities should be more formal and required.

The six CDQ groups are in competitive positions with each other regarding the state's allocation of quota. They are in a cooperative relationship with each other in participation in the North Pacific Fishery Management Council process, which affects the overall quota available and the regulation of fishing effort. The groups have attempted to form a coordinating body, the Western Alaska Fisheries Development Association, to serve as a communication forum and a way for them to synchronize their participation in the federal and state processes. This synchronization role has not been entirely successful due, in part, to the fact that the differing priorities and competition for allocations does not lend itself to coordinated efforts.

The oversight of the CDQ groups by the State of Alaska provides a way for difficulties in the management of any one of the six groups to be addressed. This feature of the CDQ program distinguishes it from the preceding economic development effort under the Alaska Native Claims Settlement Act. The fact that the CDQ groups are not individual membership corporations also distinguishes this program from the ANCSA corporations.

Since one of the motivations for establishment of the CDQ program was the poverty of the communities involved, the supervision and oversight from the State has a reason to counteract the effect of lack of business experience in the communities in the program. The leases of quota under the pollock portion of the program involves large sums of money. As might have been expected, there were some initial missteps in the sale of pollock quota, and the oversight system has worked as a way to address such problems.

The results for the supervision of investments of the returns from the lease of quota are more difficult to address, given the short time period in which these investments have been in place and the general uncertainty which exists in fisheries. The different CDQ groups have undertaken different investment strategies, and the state has not insisted on a common investment strategy. The State seeks to assure itself that minimum precautions such as the completion of due diligence investigations are taken. Each CDQ group is expected as part of due diligence to investigate the facts presented to them by potential partners to assure themselves that the information they are receiving is accurate. The danger of micro-management by the State is a possibility raised by the managers of one of the CDQ groups; the committee did not receive additional testimony or information indicating that micro-management by the state was a concern and did not assess whether or not this is a serious problem.

The committee found evidence of some social friction between native and non-native residents in villages having relatively large non-native populations. For instance, the committee received testimony relating to these frictions in Nome and Dillingham. Generally, each village has an internal governance structure that affects the opportunities of residents to participate in Community Development Plan discussions and decisions. For example, each village in the Coastal Villages Fishing Cooperative has chosen its "Traditional Council" (federally recognized tribal governments) to represent it. The choices made by villages will affect the degree to which native and non-native residents are included in the program. Given our charge and resources, this committee cannot offer a prescription for resolving these tensions.

Conclusion

The CDQ groups were given a unique governance structure that includes elements of both state and federal oversight, which is appropriate given the goals of the program. But the extensive and variable criteria used by the

**State and federal governments in allocating quota among the groups causes
decisionmaking to be inconsistent and difficult to evaluate. That the lists of
evaluation criteria are not entirely consistent with one another in either content or order of listing presents additional opportunity for confusion among
the CDQ groups and the public in evaluating the logic and fairness of the
decisions made by the governor and ratified by the Secretary of Commerce.**

Recommendations

• State and federal criteria for the allocation of quota based on performance
and plans should be less complicated than they are and should also be consistent
with one another. We recommend that changes be made to simplify the criteria,
in consultation with the CDQ groups.

• The committee notes that the criteria currently are used for two purposes:
to allocate quota equitably and to encourage good management. One way to
clarify some of the confusion created by using the criteria in this way would be to
separate these two purposes into two allocations of quota. A "foundation quota"
would address issues of equity and a "performance quota" would address issues
of performance. The foundation quota (likely more than half of the allocation)
would be allocated on measures of population, income, employment, and proximity to the fishery being allocated. The performance quota (the remainder)
would be allocated based on clearly defined performance measures such as
accomplishments of the Community Development Plan goals, compliance with
fishing regulations (e.g., regarding bycatch), quality of Community Development
Plans, and so forth.

• One way to improve responsiveness of the CDQ groups' managers to the
communities would be to improve communication. Although the idea of locating
the headquarters of the CDQ groups near potential business partners and the State
government may have made sense in the early years of the program, as it matures
and the management proves its business capability, relocation of the headquarters
to the communities may have significant benefits in terms of responsiveness to
the desires of the community members.

• Communication would be further improved if the confidentiality rules
and the rules for making information available to constituents were improved.
NMFS and the state need to collaborate to resolve any potential conflicts between
state laws regarding the confidentiality of financial data and the evaluation of the
CDQ program objectives. Information on the number of people employed by the
program and the earnings in each of the communities should be provided.

• Although some of the CDQ groups have created newsletters to communicate with their constituents, a requirement that newsletters, town meetings, or
other forms of communication appropriate to reach community members in the
region be provided might be a helpful step in improving communication in the
communities.

DEVELOPMENT OF HUMAN RESOURCES

The CDQ groups have dedicated some $4.5 million to education and training over the last five years, making this an important component of the CDQ programs in western Alaska. A closer look, however, reveals a lack of long range planning for the human resource needs within some of the CDQ organizations. Indicators include the many comments heard during site visits and also the Pete (1995) and Hensel (1997) reports that indicate there are local people unfamiliar with the CDQ program and the potential benefits it has to offer. This calls for greater attention to community outreach.

The CDQ groups have now had enough time to operate that a review of the effectiveness of the education and training components of their programs might be in order. There are cases where the training is offered at the forefront of programs to develop on-shore processors, design, and construction of boats. In addition, programs exist for maritime training for those local residents in need of this specialized certification. In most cases, however, capital expenditures are the most common achievement of the CDQ groups, with on-shore processors, harbors, docks, and necessary support facilities being constructed or purchased, but with no plan for their continued maintenance and operations. When observers ask, "what have the CDQ groups accomplished," such capital expenditures are clear answers. It is harder to document results associated with the 3,425 individuals funded for higher education scholarships or vocational/technical training.

The federal regulations for CDQ pollock, halibut, and sablefish (Section 679.30 (b) (1) (vi) of 50 C. F. R. 679.30 - 34, updated 10-07-96) mandate that the Community Development Plans include information describing the "existing fishery related infrastructure and how the Community Development Plan would use or enhance existing harvesting or processing capabilities, support facilities, and *human resources*" [emphasis added]. From the data presented to the committee, it appears that only one CDQ group furnished detailed information on the development of local human resources. This information indicates approximately 30 percent of their population has received some form of skills upgrading or training for fisheries related positions. While the bottom line of this CDQ program does not appear as profitable as others, it should be noted that they have made a conscious effort to invest in the development of their human resources as part and parcel of their fisheries infrastructure.

The strategy used by another CDQ group is to fund a scholarship trust with approximately $1.1 million from their financial portfolio. Through this scholarship program, financial aid is offered to full-time higher education students who are residents of their CDQ villages. This scholarship program was initially funded with 10 percent of pollock CDQ royalties to be earned through 1998. In looking at this strategy, there are some who view this approach as "hoarding" the funds rather than developing human resources. Others may view this approach as appropriately cautious in response to concerns about the duration of the program. Such funds could provide security if quota allocations change or the program is terminated. With the CDQ program emerging from the embryonic stage, it will be interesting to observe the progress of this approach.

All the CDQ groups have some type of scholarship fund. Some are administered through the University of Alaska. However, it is difficult to determine how the statistics are compiled for State reports. For instance, reports indicate 619 students have received higher education funding since 1993 but do not explain whether the statistics consider a student every year of attendance or once on initial enrollment. Nor does it describe which career paths students are selecting, the number of students who have successfully completed college, the number of students working in their career field, or the number working in the fishing industry.

The largest area of fishery sector employment appears to be on the fish processing line on a factory trawler, the so-called "slime line." Generally speaking, the "slime line" involves long hours in the processing facility preparing fish to be run through filleting machines. It is the tedious, dirty work assigned to low skill workers, and offers the greatest number of job opportunities; higher level opportunities so far have been limited. While construction jobs associated with capital construction projects hold potential for spin-off employment associated with the activity of the CDQ group, the data are not readily available.

Conclusion

Education, training, and other activities to develop human resources in the participating communities are an explicit part of the CDQ program mandate and a key element in ensuring the program's success because stable, healthy communities depend as much on people as on economic growth.

Recommendations

• To be truly effective, over the long-term, the CDQ groups must have education and training elements. These elements should not be haphazard, but carefully planned and coordinated so they meet long-term community needs. Both vocational training and support for higher education will help members of the community acquire the skills and knowledge needed for more advanced technical and managerial positions.

• CDQ groups need to do a better job of disseminating information that describes the educational and training opportunities open to the use of program funds. They also need to improve their recordkeeping of education and training initiatives so the results can be monitored over time. A common framework for recording and reporting their efforts would be useful.

PROGRAM DURATION

Questions have been raised about the desired duration of the CDQ program. The committee believes that the CDQ program must not be seen as a short-term solution to a long-term problem in western Alaska. Contemplation of termination of the program suggests a view of development as a terminal concept. There may be a perception in some quarters that there will come a time when the CDQ program can be declared to have achieved it goals, and be terminated. However, as discussed in Chapter 2, the purposes for which the program were created, such as long-term economic development, are not terminal concepts.

A comparison between the CDQ program and other explicitly development-oriented features of American fisheries policy in general may be illuminating regarding the suggestion that development has easily identifiable endpoints. The creation of the exclusive economic zone (EEZ) itself was a development policy. Yet those who advocate ending the CDQ program do not advocate ending the EEZ or other allocations that have benefited them. The American fishing fleet would never be content to compete with Japanese and Russian fishing vessels off the coast of Oregon or elsewhere; to do so would threaten capital investment in vessels, shore-based and off-shore processing facilities, jobs, and related infrastructure that has blossomed over the past 15-20 years.

Clearly the CDQ program is similar to other features of American fisheries policy, including allocations and the EEZ itself, in that they can be modified by Congress or the regional councils. However, singling out the CDQ program for termination because it is a "development" program strikes the committee as an odd partitioning of fisheries policy and an ill-founded conceptualization of the development process. As with other aspects of fishery policy within the EEZ, the CDQ program is an on-going program that creates new economic opportunities.

Economic development, whether it is along the Bering Sea or elsewhere, is not something that can be turned off and on. Economic development is a continual process of investing in human capital, natural capital, and technology for the betterment of families and individuals. Economic development is a continual process of change, with the essential trait being that new opportunities are pursued, new challenges are faced, and new goals are set. Economic development for communities is the structural transformation of a region and its people—more children are in school for longer periods, more jobs are available for residents of a particular place, more investments in infrastructure are undertaken, the health status of the local population improves, and per capita incomes gradually rise. However, it may be entirely appropriate to evaluate the changes that the CDQ program has created in the coastal communities and review the progress of this program in achieving improved economic opportunities and growth.

In addition to assessing the CDQ program in the Bering Sea, the committee was asked to assess the feasibility of developing some kind of CDQ-like program for the western Pacific. Again, it is difficult to undertake that task if it carries with it a presumption that at some time in the near future the program will "not be needed." This does not mean that programs that are failing must be continued. It is, rather, to point out that programs that are succeeding cannot—by that very success—then be regarded as candidates for elimination. To follow that logic is to accept a perverse notion of public policy; programs that both fail and succeed could be eliminated. Of course, some programs are of a sort that their success warrants a "sunset clause." Development, however, is a long-term process.

Conclusion

The CDQ program must be a long-term program because it deals with a long-term issue: development of healthy, sustainable communities in coastal Alaska. Long-term development requires stability in the underlying policy base so decision-makers can make choices that balance current and future needs.

Recommendations

• The original CDQ program was a three-year trial. It was subsequently extended and then made a more permanent part of the fishery management system with the passage of the Magnuson-Stevens Fishery Conservation Act reauthorization in 1996. As noted earlier, this program has been successful in bolstering community development in western Alaska. It has passed a crucial point in its evolution and we should expect the allocation of harvests to the CDQ groups to become a long-standing, if not permanent, feature of the federal fishery management system in the North Pacific.

• The committee recommends that the CDQ program should be reviewed on a periodic basis to determine if the preliminary trends observed by the committee continue in the future. Reviewing the CDQ program in another five years may provide important additional information on the effects of the program and provide valuable suggestions for modifications to its management.

ECONOMIC SUSTAINABILITY AND ENVIRONMENTAL STEWARDSHIP

The CDQ program does not now address environmental stewardship, but if the goal of the program is long-term economic sustainability in the region, it will need to give more emphasis to environmental considerations, specifically the condition of the resource base (the fisheries). All ecosystems respond to natural and anthropogenic perturbations, and even the highly productive Bering Sea ecosystem is subject to such stresses. The 1996 NRC report, *The Bering Sea Ecosystem*, concludes ". . . it seems extremely unlikely that the production of the Bering Sea Ecosystem can sustain current rates of human exploitation of the ecosystems and the population of all marine mammals and bird species that what we believe existed before human exploitation—especially modern exploitation—began." This thought seems prescient because in 1997 many fish stocks were down (whether due to climatic variation or the impacts of harvest is not clear) and thousands of birds starved and nests were abandoned because of a natural shift in the ecosystem (Baduini et al., 1998). Experience with California sardines suggest that it is almost impossible to understand the very complex interactions between natural shifts and anthropogenic impacts (McCall, 1990). This is why it is so important for management to be sufficiently conservative to ensure rapid recovery from large scale natural changes in ocean climate. Such balance is an explicit management goal in the North Pacific fishery management plans, and the North Pacific Council has tended to adopt conservative management measures when the results of a stock assessment are uncertain.

Expected variability of resource productivity and unexpected changes in the Bering Sea suggests that CDQ groups should be concerned about how they will deal with environmental surprises. Currently, the CDQ groups have a quota allocation that, while subject to some political uncertainty, seems to offer a long-term allocation of fishery resources. These groups may therefore be concerned that management practices sustain the resources for future generations. Since many of the communities are participants in subsistence economies, they generally recognize the importance of sustainable use of natural resources.

Conservation of fishery resources depends on actions taken in the North Pacific Fishery Management Council, the National Marine Fisheries Service, the Alaska State Board of Fisheries, and other entities under international treaties. But at the community level, the CDQ groups face two related policy issues: how they will structure their investment policies to deal with ecological uncertainty, and to what extent will they become involved in management to address issues of ecological sustainability?

The first issue, dealing with ecological uncertainty, is part of the general issue of pursuing economic sustainability. The idea of economic sustainability has a variety of interpretations in the literature. Some interpretations refer to the nature of the capital stock that is left for future generations. Within this general perspective, "weak" sustainability allows for the substitution of constructed capital for natural capital, while "strong" sustainability insists that the quality and quantity of natural capital must be retained at all cost. Other economists discuss sustainability in terms of the welfare to successive generations into the future. Here, sustainability is interpreted as a situation in which human welfare does not decline over time. Another main theme in the literature on economic sustainability is coping with uncertainty. The flows of products to economic systems from ecological systems are highly variable. A sustainable economy, in addition to maintaining stocks of natural, human, and physical capital, must be able to cope with uncertainty. It is also important to consider that markets for fishery products can be particularly difficult to predict, including changes in the demographic situation in the communities. For the CDQ groups, biological and economic uncertainties are augmented by political uncertainties. The CDQ groups do not have a choice about what their share of the natural capital will be, since their quotas in the several fisheries are already determined by the Council and the State. The CDQ groups do have options about their levels of constructed capital, human capital, and other types of investments. Diversification would seem to be a prudent investment strategy.

Regarding the second issue, management involvement, increased attention to long-term sustainability is evident throughout the sector. For instance, significant changes in the 1996 re-authorization of the Magnuson–Stevens Fishery Conservation and Management Act have emphasized conservation. Three of these changes emphasize (1) ecosystem-based management, (2) protection of essential fish habitats, and (3) avoidance of bycatch. The CDQ groups can choose to become involved in all of these issues. Ecosystem-based management is fraught with semantic problems and lack of data. The most useful approach might be to admit that adequate data will never be available to fully understand marine ecosystems, and instead fisheries managers should focus on precautionary management strategies that avoid risk to the long-term sustainability of the system. To follow this approach, the CDQ groups should consider becoming interested in total allowable catch determinations, especially with regard to assuring that the exploitation does not reduce any stocks or any food web relationships below thresholds from which recovery does not take more than one fish generation. Because of the structure of the CDQ program, the CDQ groups are not directly involved in setting management goals related to resource conservation. They do have some form of representation on the council and within the council's industry advisory panel. The long-term assurance of exclusive rights to the resource that the CDQ program provides may create an incentive for the CDQ groups to conservatively manage the fishery. Although the CDQ groups have not developed explicit plans to develop conservation measures, further efforts to ensure the long-term sustainability of the resource could benefit the CDQ groups in the long-term.

Those managing the CDQ groups should be concerned about the relationship between natural environmental change and change resulting from fishing. To understand that relationship, a long-term research program is needed, including monitoring protected reserves or sanctuaries in areas large enough to evaluate natural fluctuations independent of fishing-induced changes. If this activity is undertaken, possibly by the National Marine Fisheries Service, representative reserves may be established in areas close to communities that might otherwise be open to CDQ fishing.

Protection of essential fish habitat is of obvious concern in regions where bottom trawling is utilized. There is evidence that trawling can have massive effects on many benthic habitats as well as taking a considerable bycatch of both commercial and non-commercial species. In recognition of this there are no-trawl zones designed to protect the crab fisheries. It is relevant to note that the crab fisheries themselves can have environmental impacts from bycatch and especially from continued "ghost" fishing by lost traps which attract predators such as octopus, fish, and snails even if the panels do open so that the crabs can escape.

The Magnuson–Stevens Act specifically addresses broad concerns about bycatch. Bycatch occurs not just from bottom trawling and trap fishing, but all fisheries have some bycatch which potentially impacts other populations in the ecosystem. CDQ groups will need to comply with the Magnuson-Stevens Act and the National Standards for Fishery Management objectives of reducing bycatch; such compliance should contribute to healthier fisheries which will benefit CDQ groups in the long term.

Conclusion

Economic sustainability implies programs and policies that offer the greatest assurance of economic options over the long-term to a population that chooses to remain in specific locations. That is, given alternative economic futures for a people (or for a community), economic sustainability would entail choosing that future with the lowest probability of inducing economic decline as measured by a range of indicators. Economic sustainability is but one part of the larger problem of ecological and socio-cultural sustainability. Clearly, communities that squander their local environmental resources (or that fail to maintain cultural and social processes and structures) will be incapable of economic sustainability. Large-scale commercial fishing activities can have negative impacts on ecosystems, either independently or through interaction with natural fluctuations. Because the CDQ program is designed specifically to increase participation in fisheries activities and at the same time improve the long-term economic conditions of the participating communities, special emphasis should be given to environmental stewardship.

Recommendations

• Overall, concern for the long-term health of the Bering Sea ecosystem needs to feature more prominently in the CDQ program. Local level concerns about environmental sustainability and stewardship need to be able to be expressed in a meaningful way throughout the management structure, beginning with effective communication of local concerns to the CDQ group management and continuing on up through the Council process. The allocation process can foster enhanced emphasis on environmental sustainability by directly recognizing such emphasis when applications from the various CDQ groups are evaluated.

• Economic sustainability is dependent upon sound environmental stewardship. In order for the CDQ program to help build a sustainable economy in the region, it is imperative that the underlying resource base—the fisheries—be used in ways that are sustainable over the long-term. This will require explicit, indepth, continuing analysis of the condition or health of the fishery resource and management that can respond and adapt to changes in this condition.

5

Broader Issues and Considerations

This committee was asked to comment on the potential applicability of the Community Development Quota (CDQ) concept to areas beyond western Alaska, specifically whether the concept might be used in the western Pacific. Both the relative novelty of the CDQ concept in fisheries management and the initial promise of the western Alaska CDQ program contribute to this interest in the broader application of the CDQ concept. At the broadest level, what is of interest is the general idea of community allocations and also community management.

What emerges from a review of the western Alaska CDQ program is an appreciation for how specifically that program was tailored to the region in which it was applied. The very specificity of the Alaska program is useful in identifying critical characteristics, or design features, that must be addressed by anyone exploring the possible application of the more generic CDQ concept. The CDQ program offers insight into the interaction of state and local managers, the use of quota-based management, and the role of oversight in community-based development. These components, and other components of the CDQ program discussed in this chapter, are likely to be important when considering any community-based approach to development and fisheries management. Thinking about the broader applications of CDQ-style programs also addresses many issues of interest to Congress regarding harvest privileges. The following sections examine specific features of the western Alaska CDQ program for both their implications in the context of the Alaska program and for illustrations of more generic issues.

SOME INFLUENTIAL CHARACTERISTICS

The western Alaska CDQ program is based on four critical concepts: quotas, communities, corporations, and oversight. These three concepts are familiar to participants in the North Pacific Fishery Management Council process and are reviewed briefly here.

Quotas

Quotas are the most fundamental aspect of the CDQ program. Quota management has different meanings throughout the world of fisheries management. In the context of the North Pacific, quotas are a derivative of total allowable catch (TAC) management. TAC management has a long history in the North Pacific region and is almost uniformly accepted as the legitimate underpinning of subsequent management measures. For those wishing to follow the specific example set by the Alaska CDQ program, a critical feature is the existence and acceptance of TAC management. In contrast, in some regions introduction of TAC management itself is likely to be as controversial. Conceivably, in other regions not using a TAC alternative management, techniques such as permit allocations and restrictions, special areas, gear restrictions, or limits on the number of fishing days could be used to allocate a portion of a fishery to a CDQ-like program.

The fact that the CDQ program is based in TAC management conferred an important element of legitimacy to the program by grounding it within the Council process. TAC management provides the basis for a wide manner of allocations that have been employed by the Council in addition to the collective quotas of the CDQ program, including, for example, the individual fishing quotas (IFQs) for halibut and sablefish and the inshore/offshore allocation in the Bering Sea pollock fishery. Like other TAC-based allocations adopted by the Council, the CDQ program was designed to bestow economic opportunity, not to challenge the Council's management authority and functions.

Basing the CDQ program on the familiar grounds of a sub-allocation of the TAC provides more, however, than just a measure of legitimacy within the larger political dynamics of the North Pacific Fishery Management Council process. The bestowal of economic opportunity provides a characteristic focus on access to the economic benefits available through resource exploitation. The western Alaska CDQ program was designed to achieve a goal of inclusion. The coastal communities of western Alaska were perceived to be excluded from the ongoing development of the commercial fisheries in the Bering Sea and Aleutian Islands. Further, there was an apprehension that specific management allocations would institutionalize and perhaps exacerbate existing patterns of acute underdevelopment and marginalization of the local population. The form of inclusion fashioned under the Alaska CDQ program is one in which the CDQ groups receive a harvesting privilege and control of any economic benefits associated with that privilege.

Those interested in future applications of the CDQ concept should note just how specifically the North Pacific Fishery Management Council designed the nature of this form of inclusion. Sharing in economic benefits is not the same as, for example, sharing in management responsibilities. One can imagine a program with different mechanisms of inclusion such as a direct allocation of voting seats on the Council, direct control over setting the TAC itself, or even some form of autonomous participation in the subject fisheries completely outside the Council process. In contrast, the western Alaska CDQ program appears as a program crafted along much more familiar lines that bestowed economic opportunity while remaining within the overall framework of the Council.

Communities

Much like the acceptance of TAC/quota management, the existence of a commonly accepted view of "community" contributed to the prompt initiation of the western Alaska CDQ program. The relative ease with which target communities were identified in the Alaska case may tend to obscure an important point in terms of the broader interest in CDQ-style programs. Implementation of a community development program requires a definition of "community" and the ability to direct program benefits to the defined community. The concept of community involves both the idea of a geographic community (a community of place) and a social community (a community of interests). To the extent that there is a lack of correspondence between target interests and target places, substitution of one form of community for the other in the design of a program may be problematic.

The western Alaska CDQ program is defined in terms of 57 geographic communities. Any potential discord with the programmatic interest in providing benefits to the Alaska Native peoples of the Bering Sea and Aleutian Islands is minimized because of the nature of the communities themselves and because of internal and external recognition of these communities. The Alaska CDQ communities are overwhelmingly small villages of Native Alaskans with a strong sense of community identity and with well-recognized problems of development. From an external perspective, the Alaska Native villages are recognized as such in federal law. These factors make a geographic community basis seem like a natural choice in the western Alaska context. As noted earlier in this report, tensions are present in the Alaska program within those communities with significant non-native populations and ambiguities arise over who are the eligible residents within the eligible communities defined in the federal regulations.

The Alaska CDQ program suggests an obvious lesson for architects of any future CDQ program: framing a program in terms of geographic communities will be most effective when these communities are, or are nearly, synonymous with the intended beneficiaries. Where the intended beneficiaries are dispersed among relatively large heterogeneous populations, a geographic community basis is likely to be unsuccessful (as for example when a target population is effectively lost among the general population of a large metropolitan area).

While defining the program in terms of geographic communities works because of the specific circumstances of western Alaska, this same context poses a challenge. The 57 communities are extremely isolated (usually even from one another) and are broadly dispersed in geographic and cultural terms. This dispersion presents a logistical problem which was addressed in the Alaska program through the creation of private corporations.

Corporations

A third characteristic of the Alaska CDQ program is organization around nonprofit corporations representing collectives of the eligible communities. Like using quotas as a key element in the program, the mandatory basis in a corporate organizational form was partly a decision based on familiarity, in this case familiarity with the regional and local native corporations formed in response to Alaska Native Claims Settlement Act (ANCSA). The experiences of Native Alaskans with the corporate organizational form have been both positive and negative, and the negative experiences have influenced some CDQ group leaders to approach the CDQ program with a measure of fiscal conservatism. The experiences of Native Alaskans with the corporate organizational form gives them knowledge that may not be present in other areas where CDQ-like programs may be of interest. Even with this prior experience, the learning curve is substantial and, as noted in Chapter 4, the education of the CDQ groups in corporate management continues.

The corporate model contains an inherent challenge to the emphasis on community development: the corporation is not the community. Although the committee did not see evidence of this in the current groups, diversion of program benefits to the corporation may introduce new and perhaps unintended beneficiaries. In the Alaska CDQ program, the corporate organizing model has helped address the logistical hurdle posed by the extreme dispersion of the CDQ communities. The corporate organizational form has also complemented an initial focus on capital accumulation necessary to sustain a variety of development activities over the long term. Still, the committee observed that community residents are sensitive to the distinction between community development and corporation management and there is some tension between the villages and the corporate entities that constitute the CDQ groups.

It is important to distinguish the various types of corporate structure that may be used. In the ANCSA corporate models, the stakeholders are a fixed group of beneficiaries who retain their vote and benefits wherever they move. In the case of the CDQ corporate structure, the groups are operated by an elected board from the designated communities. These corporate models can be expected to produce differing interactions between the villages and the corporate entities. The Alaska CDQ program may be at a crossroads in respect to the balance between attention to developing a strong corporate structure and community development. Heavy emphasis on development of a strong corporate structure may be appropriate during the start-up phase, but as time passes and the program matures, that kind of emphasis on the corporation as beneficiary can become entrenched to the detriment and displacement of the originally intended beneficiaries.

For future CDQ-like programs, two lessons stand out. First, the corporate governance structure may not be as familiar, efficient, or as relatively easy to adopt as in Alaska. Second, there may be a delicate balance between governance structures operating at one level (the corporation) and program goals targeted at a different level (the community). The balance struck between the development achieved at these different levels may ultimately affect what it means for a community development program to be "successful."

Oversight

A final characteristic feature of the western Alaska CDQ program is the formal oversight role assigned to the state of Alaska and the National Marine Fisheries Service. The distinctiveness of the oversight role in the Alaska CDQ program is illustrated by comparison to other allocations effected by the North Pacific Fishery Management Council—other allocations are not accompanied by a formal oversight role. A CDQ-style program could be fashioned without an oversight role, thus one question confronting those interested in possible application of the CDQ concept is whether to mimic the oversight functions of the Alaska program.

Experience with the Alaska program suggests that decisions about the appropriate level of oversight may be challenging. On one hand, the committee repeatedly heard from the CDQ groups that "state oversight is a good thing." Among the benefits attributed to external oversight were the sense of a level playing field among the six CDQ groups and protection against bad investments that might tarnish the entire program let alone the particular group involved in such an investment. The presence of the oversight role also seems to have significantly contributed to the legitimacy of the program, particularly in the beginning when the program was perceived as a radical creation and was more controversial than it is now.

On the other hand, as has been discussed in the previous chapter, oversight (or the carrying out of oversight functions) is a continuing source of irritation and confusion within the program. The challenge that the Alaska CDQ program poses is that oversight appears as beneficial, appropriate, irksome, and puzzling all at the same time. Determining what kind and how much oversight is appropriate is a significant balancing act. The subsequent carrying out of such oversight in a consistent and coherent manner requires further skill and attention. Finally, it should not be expected that the balance point will remain stationary as the program matures. These challenges will continue to confront the Alaska CDQ program and any future application of the CDQ concept.

RULES OF INCLUSION AND EXCLUSION

In any resource management regime based on the allocation of property or privileges, the rules of inclusion and exclusion are likely to be controversial to design and implement. The Alaska CDQ program has experienced fewer controversies associated with the initial allocation of privileges than the individual fishing quota program in the halibut and sablefish fisheries off Alaska. However, the CDQ program is clearly not free of the vexing distributional issues associated with determinations of who is included and who is excluded.

In the western Alaska CDQ program rules of inclusion and exclusion are determined at two distinct levels, the local level and the overall program level. At the local level, rules governing the distribution of benefits and opportunities are addressed by the internal governance structures of the CDQ corporations and the villages themselves. Again, the relationship between the villages and the CDQ corporations present some instructive issues for those interested in CDQ-like programs. On one hand, the role played by the CDQ corporations in decisions affecting the local distribution of benefits (including participation opportunities) suggests an incomplete devolution of authority to the local level (a distinction of critical relevance to any interest in community management). On the other hand, the leadership of the villages frequently overlaps the leadership of the CDQ corporations. This can be advantageous, but also raises some concerns about separation of responsibilities and opportunities for broad participation in the CDQ organizations.

Whether local level decisions are truly determined locally or by some proxy, one remaining issue is the potential problem of domination by a local elite. Political power is rarely, if ever, evenly distributed across a community of any size. Complaints inevitably arise that the consequences of allocation decisions are not evenly distributed across the spectrum of affected participants nor is access to the decision-making process equal. However, mere devolution of authority over allocation decisions does not in itself resolve the potential problem of distributional inequities. Rather, the problem is simply shifted to the local level. Communities can be highly factional, small communities perhaps particularly so, which can magnify disparities in the distribution of political power. Further friction can arise in a governance structure involving a collective of communities if one community has, or appears to have, more voice than others.

Although many of the Alaska CDQ villages are characterized by factionalism, it was not apparent to the committee that this has led to substantial problems in the western Alaska CDQ program. One area where the problem of factionalism may emerge concerns the distribution of fishing opportunities in CDQ fisheries that are conducted locally (or could be conducted locally) such as the halibut fishery. To date, these fisheries have been conducted as local open access fisheries and the communities and the CDQ groups are wrestling with the challenges of allocating these participation opportunities. Daily trip limits, IFQ-like individual total catch limits, and full-time residency requirements have all been explored or used in allocating access to the CDQ halibut fisheries. As demand for local participation opportunities grows, one can expect the potential for intra- and inter-community tension to increase and the issue of control over local access policies to become more critical.

At the overall program level, significant rules of inclusion and exclusion are established in the authorizing federal regulations in the form of criteria for determining eligible communities (criteria that have now been incorporated into the Magnuson-Stevens Act). For example, the community eligibility criteria state that an eligible community must be within 50 miles of the Bering Sea coast and cannot be located along the Gulf of Alaska. Both of these criteria create "edge-effects" in the form of inter-community friction as communities are divided by boundary lines into haves- and have-nots. Some of the CDQ groups have tried to address these effects by opening their training and scholarship programs to the residents of neighboring villages beyond the 50 mile boundary, but capital projects may not be undertaken in these villages and they have no direct role in the governance of the proximate CDQ group.

The exact rationale behind the particular geographic limitations on the Alaska CDQ program is not clear. The committee received testimony suggesting that the program's design was related to a unique combination of poverty, underdevelopment, lack of established fisheries infrastructure, and indigenous populations. With particular reference to the program's geographic boundaries, this uniqueness is questionable. There are clearly other communities with similar profiles that lie outside the program's boundaries. Review of the historical record associated with the program suggests that the program's boundaries reflect political dynamics within both Alaska and the North Pacific Fishery Management Council. For example, the 50-mile boundary rule was originally conceived as a 30-mile boundary rule and changed to 50 miles during Council deliberations. Institutional structures also influenced program boundaries as evidenced by the exclusion of all coastal communities north of 66°N latitude for the simple reason that the Council's fishery management plan for the Bering Sea and Aleutian Islands terminates at that latitude. The Council has no fishery management plan for the Chukchi Sea region, thus no active authority in the region and no ability to apply the CDQ program in the region (Oliver, 1993).

Examination of the historical and traditional patterns of fishing and trade activities by the region's communities reveals that the rule requiring CDQ communities to be within 50 miles of the Bering Sea has implications for patterns of traditional relationships among communities in different regions of western Alaska. For instance, the application of the 50-mile rule in the Bristol Bay region prevented at least 10 communities, all ANCSA villages with Alaska Native residents, from CDQ membership Some of these communities have participated in salmon and herring fisheries in Bristol Bay, in many cases since World War II. This has created a new division of communities not based on their previous commercial fishing histories and cultural practices but by an arbitrarily imposed geographic boundary.

The limited geographic application of CDQs in Alaska presents some interesting challenges to those interested in broader application of the CDQ concept. The specific geographic boundaries of the Alaska program serve to emphasize how particular that program is and how it was fashioned in the specific context of the politics of fisheries in Alaska and the North Pacific Fishery Management Council. Others may be interested in the application of CDQ-style programs on a more expansive basis. For example, CDQs hold some appeal as a fishery management tool independent of their utility as a development tool. Those interested in broader use would need to address how eligible communities be selected and what the program boundaries would look like. This question addresses the subject of threshold criteria to be used in determining eligibility to a program.

The threshold criteria for determining communities suitable for a CDQ-style program depend in large part on the nature of the program itself. It is not possible to establish threshold criteria for some hypothetical situation devoid of both context and intent. Assuming that a geographic community basis even makes sense in the first place, and that the emphasis is on community development for indigenous peoples, and that one wishes to address the distributional equity aspects of previous fishery development efforts, then community eligibility criteria similar to those employed in the Alaska program may be appropriate.

In contrast, a program focused on community allocations as a fishery management tool (perhaps even applied to the entire TAC) might feature a relatively unrestricted pool of eligible communities. The Alaska CDQ program does not really offer much instruction to this latter type of program, except to present another challenge: could a CDQ program be applied to coastal communities throughout a state rather than just to a particular portion of a state?

In response to a comment in the final rule for the original pollock CDQ program, the National Marine Fisheries Service articulated an interesting legal theory associated with the limited geographic range of the Alaska CDQ program:

> [T]he CDQ program does not discriminate between Alaskans and non-Alaskans on the basis of residency. The impact of the CDQ program in setting aside a pollock reserve for use by western Alaska communities . . . falls equally upon similarly situated Alaskans and non-Alaskans. Regulations that are determined to discriminate among residents of different states, based on their residence, would not be approved. (FR vol. 57, No. 248:61329, 1992)

Again, the Alaska CDQ program appears as a very particular construction of what could be a very general concept, this time in terms of the program level rules of inclusion and exclusion. In the 1996 reauthorization of the Magnuson-Stevens Fishery Conservation and Management Act, Congress placed a moratorium on consideration by the Council of any change to these rules. The clash between Congressional action and what appears to be growing interest in expansion of the CDQ program will no doubt be mediated in the normal political processes affecting fisheries in the North Pacific.

One area in the North Pacific that has been proposed to be included if the CDQ concept were to be extended to other areas within Alaska would be communities in the Gulf of Alaska. Two possible ways to extend the CDQ concept into the Gulf of Alaska have been suggested. The first approach focuses on providing potential new CDQ groups in the Gulf of Alaska with Bering Sea pollock quota because the Gulf of Alaska does not have a comparable unallocated stock of size and value. This approach might be problematic. First, it severs the philosophical association of quota with access to the collective resources of a proximate marine ecosystem. Second, accommodating any new program by increasing the current 7.5 percent overall CDQ allocation would face substantial industry objections. Third, redirection of a portion of the existing 7.5 percent CDQ pollock allocation would face substantial industry objections. Finally, redirection of a portion of the existing 7.5 percent CDQ pollock allocation to new CDQ groups/communities would result in diminishment of benefits to western Alaska and also provoke resistance. These latter two problems are testimony to the potent sense of entitlement associated with fisheries allocations.

A second suggested possible way to extend the CDQ concept to the Gulf of Alaska and southeast Alaska focuses on the IFQ halibut and sablefish fisheries of the Gulf. These suggestions have been made recently by a number of organizations representing Native Alaskan interests in the region. Because both halibut and sablefish stocks in the region have been fully allocated into an IFQ program, this proposed route would be somewhat different than the current CDQ program. Currently, regulations in the halibut and sablefish IFQ program limiting the amount of quota share that any one entity can hold restrict the ability of CDQ-like groups to form and hold quota. These rules may have to be modified if communities were to hold quota share. It would authorize the creation of CDQ-like groups, comprised of one or more communities. This would not require a reallocation by the Council to the new groups; rather, these proposed groups would be authorized to purchase IFQ from present individual owners. This quota could then be leased or allocated to group members thereby creating new fishing opportunities in these communities.

This second route could preserve the linkage between communities and nearby resources and requires a significant degree of organization and entrepreneurial commitment be demonstrated to achieve group status. Given the evident underdevelopment difficulties of many smaller Gulf of Alaska and southeast Alaska communities, the possibility of creating some kind of CDQ-like organization might be worth additional consideration within the normal framework of the

North Pacific Fishery Management Council process. The committee did not evaluate the historic dependence on fisheries in these communities, the potential boundaries and criteria that could be used to define those communities, or the level of development, employment, or alternatives to fishing in the communities where the CDQ program is being considered, but the state might wish to do so before any expansion is planned. Recently, some efforts have been made to examine the holdings of limited access permits and IFQs in Gulf of Alaska communities, but the committee did not evaluate those findings or their potential implications for the expansion of the CDQ program (CFEC, 1998). It should be noted that this proposal is controversial and there is some concern that extending the authority of communities to hold quota could create a market advantages that could be disadvantageous to existing IFQ holders.

These potential impediments to the implementation of a new CDQ-like program point to a critical area of study in the realm of right-based fishing regimes. The IFQ and CDQ programs are related and those interested in expanding the CDQ concept must understand it *and* the IFQ concept.

CDQ/IFQ RELATIONSHIPS

The sequence of adoption of IFQ and CDQ programs has implications for the probability of success. If the TAC is completely assigned via an IFQ program, there is little opportunity for a CDQ program. But on the other hand, CDQ programs use only a portion of TAC, not the entire amount available, so there is room for subsequent adoption of an IFQ program. The CDQ programs in Alaska were introduced either prior to or contemporaneous with the introduction of the IFQ program. In contrast, the IFQ program for the sablefish and halibut fisheries in the Gulf of Alaska allocated 100 percent of the available TAC. Expectations associated with the IFQ program are already strongly entrenched and subsequent reallocation of the TAC away from IFQs to CDQs would be politically contentious, if not impossible. Whether or not an IFQ program based on a partial assignment of the TAC is feasible (leaving flexibility for future adoption of a CDQ program) is a question we will leave to our companion study committee.[1] But careful attention should be given to the potential for sequentially binding, and perhaps non-reciprocal, relationships between CDQ and IFQ programs. Given the nature of the two approaches to allocation, it would seem that CDQs may precede but not follow IFQs, at least in the political landscape associated with commercial fisheries off Alaska.

One possible resolution to the sequencing problem evident in the Gulf of

---

[1]A separate National Research Council committee examined the performance and effectiveness of individual fishing quotas in the United States and broader questions about quota-based allocation of fishery resources in the report, *Sharing the Fish: Toward a National Policy on Individual Fishing Quotas* (1999).

Alaska halibut and sablefish fisheries is to convert IFQs to CDQs. Rather than reassigning TAC from the existing IFQ program to a new CDQ program, IFQs could be converted into CDQs through market transactions. Communities could combine to purchase IFQs for the purpose of conversion to a CDQ. Once created, this CDQ could be treated as a royalty engine for capital accumulation, as a participation block for local fishers, or could be divided into individual allocations on a locally determined loan basis similar to options being pursued in the existing western Alaska CDQ program. Community eligibility criteria could be devised and oversight mechanisms instituted. Sources of initial capital for IFQ purchases would need to be identified and existing regulatory constraints on purchases by non-initial recipients of quota shares (which are designed to limit corporate ownership of quota shares) would need to be relaxed to permit community ownership of quota shares.

Another possible means of incorporating community-based concerns in an IFQ program is to allow communities to pool capital resources to purchase access rights (IFQs) that could then be used by members of the community. The role of the community in this approach differs from the conversion of existing IFQs into a CDQ-like program. However, under this pooling of capital resources, the community would maintain control over the use of the quota rather than a fishing vessel owner or specific crew member in the community. This approach would require clear guidelines to establish who within the community could use the IFQ, how the IFQ would be leased to those community members, as well as modifications in the existing halibut and sablefish programs to allow communities to purchase, hold, and lease quota to members in its community.

What are the distinctions between these two types of quota-based fishing regimes—the CDQ and IFQ— that might lead to a choice of one over the other? There are several important distinctions. The principal difference concerns the role of place: IFQs are nongeographic while CDQs are fundamentally geographic. By definition, a CDQ program contains a central emphasis on a geographic location. In contrast, geographic location is not a consideration in the conventional theory of IFQs. Mobility is to IFQs what the sense of place is to CDQs. Communities (in the geographic sense) are the conceptual heart of the CDQ program.

One additional difference between CDQ and IFQ programs concerns who receives the windfall associated with the initial allocation of harvest-access opportunities. In the IFQ program a subset of individual fishing firms receives a valuable asset in the form of current and future expected harvesting outcomes. The original recipient can retain this harvesting outcome or sell it. This means that new entrants into an IFQ fishery must "buy their way in"—reinforcing the extent to which the IFQ program represents a windfall to the original recipients of potential harvesting opportunities.

In the CDQ program the asset value accrues to the corporate entity (the CDQ group) which then manages it and allocates the resultant income stream to members of the local community. There is no opportunity for the CDQ group to alienate (sell) the asset and its expected income stream as in the IFQ program. New entrants into the CDQ program may not purchase their access from existing groups but would need to be allocated a share of the total CDQ harvest by the State of Alaska and the National Marine Fisheries Service. This share for new entrants would likely reduce the harvest available for current CDQ groups. In this sense, the CDQ allocation represents an income stream to recipient CDQ groups and communities but it is not a marketable asset in the same sense as the IFQ.

Greater intergenerational equity may be attained under a CDQ program because it preserves future options for the community at-large. The emphasis on the community over the individual carries with it a recognition that future generations have a claim on present assets. The CDQ program, unlike an IFQ program, prohibits the sale of the future stream of economic benefits from the quota to other potential participants. In this way, the CDQ program strengthens the claim in the fishery of future generations of community residents.

Vesting a community with specific access to fishery resources permits that community (or collective of communities) to react to circumstances that are perceived as common problems. Community goals and problem definitions can change as local circumstances change. For instance, population growth is one of the changing circumstances that often requires attention, as are shifts in economic opportunities within the community. A CDQ program can provide a source of capital for addressing changing circumstances. Another form of intergenerational adaptability inherent in the CDQ program is that a community can change the contemporary rules of access to the resource according to community conditions. These forms of adaptability mean that a CDQ-style program provides a dynamic mechanism for a community to pursue the community's conception of "development." Given that many CDQ communities require time and expenditures to learn to operate profitably, the CDQ program provides a way to assemble capital and experience that gives participants a better position in the fisheries sector. A CDQ program can enlist the services of the community's best leaders to work for the community's benefit, rather than toward private accumulation of wealth.

Various transferability restrictions could be placed on an IFQ program to address intergenerational equity issues and the loss of community access. However, such restrictions either move an IFQ program toward the CDQ end of what is an IFQ/CDQ continuum, or fail to match the ability of a CDQ program to ameliorate these community and equity concerns. The idea of an IFQ/CDQ continuum warrants closer attention than it has received in public debate over harvest privileges and quota-based fishing regimes. For example, at what point would a CDQ program be more appropriate than a heavily restricted IFQ program if community and equity concerns are strong? Conversely, when would an IFQ program make more sense than a relatively unregulated CDQ program if emphasis on the business environment of individual firms is paramount? These are rhetorical questions, but they underscore critical areas for consideration during the design stages of future programs.

The differing emphasis on place and intergenerational equity suggests a basis for another distinction between CDQs and IFQs that should be considered during discussions of transforming one type of program into the other. The goals of the CDQ program—economic and social development in communities—would not be advanced by transformation of CDQs into IFQs. This assessment reflects both the conditions that prevail in western Alaska communities and the fundamental distinction between a community and an individual.

Simultaneous exploration of the conversion of IFQs to CDQs and rejection of the reverse may provoke a charge of paternalism. But the charge only holds if centered on the interests of the present generation of individuals. If place and the claim of future generations are recognized as legitimate policy factors, then the inalienability of CDQs is a virtue, not a vice. Thus, if place and intergenerational equity are no longer policy concerns, then the transformation of CDQs should be contemplated.

Finally, there is one additional aspect of the CDQ/IFQ relationship worth noting for future program designers. As evident in the royalty fisheries of the western Alaska CDQ program, there is no need to formally transform a CDQ into an IFQ to gain some of the operational benefits offered by an IFQ program. CDQs may be treated as IFQs for an interim period of time while the community/ governing body retains the ultimate control over the underlying allocation. In addition to this critical retention of control (and thus maintenance of the inalienable character of resource privileges involved), the ability to conduct a shadow IFQ fishery within an overall CDQ program presents the possibility of semi-public collection of economic benefits and economically efficient auction arrangements. Public collection of benefits (via auctions and other mechanisms) is currently prohibited under the Magnuson-Stevens Fishery Conservation and Management Act. But there is no prohibition of the private collection of benefits. The CDQ corporations capture fishery resource benefits just like other corporate entities capture economic returns under the IFQ program. The promise of the CDQ program, however, is that the CDQ corporations are not just like any other private corporation in that there is a specific tie back to the communities (a larger public) through the program structure.

CO-MANAGEMENT/COMMUNITY MANAGEMENT

In the field of natural resources management, there is much interest in co-management and community management regimes—that is, management structures where communities have a direct role in decisionmaking about management of the resource, such as timing of fishing seasons, amount of available catch, and similar issues. The western Alaska CDQ program should be carefully distinguished from these exercises in local management authority. If "management" is understood as management of the resource (not business management), then the Alaska CDQ program is not co-management (sharing of management with a higher governmental authority) and not yet community management (full devolution of resource management authority). The CDQ program assigns rights to economic benefits via a quota share of the TAC but there is no assignment of resource management authority.

Some have argued that the CDQ groups have a management role in the coming multispecies CDQ fisheries in that they are expected to stop fishing when they have attained their prescribed harvest levels. Reliance on the holder of the harvest privilege to stop fishing (rather than relying on removal from the fishing grounds by federal agents) is the same mechanism that is employed in the IFQ program for the halibut and sablefish fisheries, and no one would seriously suggest that the IFQ program is an exercise in co- or community management. As noted earlier, the CDQ program was crafted to operate within the existing resource management structure of the North Pacific Fishery Management Council.

However, a management role could evolve in the CDQ program by two pos-

Commercial fishing is a complex and specialized activity, with the tools and techniques adapted to specific conditions. This small temporary dock and small boats are typical of the kinds used for near-shore fishing, such as seen here at St. Paul Island. What emerges overall from a review of the CDQ program is an appreciation that the program has been designed to suit very particular circumstances. Wholesale importation of the Alaska CDQ program to other locales is likely to be unsuccessful unless the local context and goals are similar to those in Alaska. (Photo by Ron Trosper.)

sible routes. The first route lies in the adjustment of the existing management regime to the complex catch accounting demands presented in the multispecies CDQ fisheries. Individual vessel level catch accounting on a species-by-species basis presents a severe challenge to existing management procedures and the CDQ groups are already being called on to devise new methods of catch measurement and reporting. Given current federal and state budgetary climates, it is conceivable that the CDQ groups will increasingly be relied upon to fill resource management needs such as being involved in catch measurement and reporting.

The second possible evolutionary route to a management role for the CDQ program is through the Council process itself. The CDQ program would not be the first time that economic empowerment led to political empowerment. While the CDQ program did not directly confer any representation in the Council process, there has been a significant evolution in the Council structure since the inception of the CDQ program. Currently, CDQ group representatives and CDQ group harvesting partners occupy 4 of the 11 voting seats on the Council and 7 out of 23 seats on the industry Advisory Panel. In essence, co-management of the management process could lead to co-management of the resource.

SUMMARY

What emerges from a review of the western Alaska CDQ program is an appreciation that the program represents an example of a broad concept adapted to very particular circumstances. These circumstances include the existing embrace of TAC management, an existing management structure with widespread legitimacy, existing community definitions, and existing familiarity with the corporate organizational form. In light of these circumstances, the Alaska CDQ program seems to be quite well tailored to the local context.

Others interested in the application of CDQ-style programs may have different aspirations and different contexts. Community allocations and community management may be appropriate as fishery management tools quite apart from their utility in contexts associated with development goals and indigenous peoples. In the broader debate over harvest privileges and quota-based fishing regimes, fishery management aims may not necessarily be synonymous with social and economic development aims. In this broader context, CDQ-style programs might be important policy options wherever issues of community and intergenerational equity are important.

As future programs are contemplated, it will be important to understand that the Alaska CDQ program is not a general model. The Alaska CDQ program is a very particular application of a largely unexplored, much more general concept. Wholesale importation of the Alaska CDQ program to other locales is likely to be unsuccessful unless the local context and goals are similar to those in Alaska.

6

Communities and Fisheries of the Western Pacific

The Alaska Community Development Quota (CDQ) program model is a specific program tailored to conditions existing in the communities and fisheries in western Alaska. In Alaska, there were clearly definable communities with clearly definable economic needs and limited economic opportunities that had been largely excluded from the fishery. In addition, the fishery was already managed by quota and a portion of the quota was being held in reserve.[1] For the purposes of this section, the western Pacific refers to Hawaii and those U.S. territories and possessions under the jurisdiction of the Western Pacific Fishery Management Council (WPFMC).

In the western Pacific, the setting and communities differ. The major differences between the fisheries and communities of the western Pacific region and the North Pacific region are: the general lack of management by quota or total allowable catch (TAC); the pelagic nature of the valuable fisheries; and the lack of clear geographically definable "native" communities in most parts of the region. Pelagic fisheries in the western Pacific are not managed by quota because the targeted species, tuna and swordfish, are highly migratory and quota management is quickly complicated by these factors. Stock assessments on which to base a TAC for highly migratory pelagic species are difficult due to the migratory

---

[1]The National Marine Fisheries Service (NMFS) sets the total allowable catch (TAC) for the fisheries in the Bering Sea. They had held 15 percent of the quota in a given year in reserve against the possibility of poor conditions in the fisheries although ultimately this was distributed at the end of the season based on how the harvest progressed. When the CDQ program was established, 7.5 percent, or half of the biological reserve, was allocated to the CDQ communities.

nature of the stock and the incomplete collection of data throughout their range. To impose a quota on domestic fishermen who fish both inside and outside of the exclusive economic zone (EEZ), as is the case with the Hawaii-based longline fishery, would disadvantage them relative to foreign fishermen. With the exception of the relatively small Northwest Hawaiian Islands lobster fishery and the inactive precious coral fishery, the fisheries of the western Pacific are managed as open access fisheries or operate under limited entry programs. The most valuable fisheries are those targeting tunas and associated pelagic species, and there is significant value in processing and transshipment. With the exception of American Samoa, and possibly Rota and Tinian in the Commonwealth of the Northern Marianas, the native peoples of the region are dispersed as minorities in larger urban and periurban communities. Thus it would be very difficult to transfer the Alaska CDQ program as currently structured directly to the western Pacific.

Increasing the participation of native fishermen in the fishery has been a long term goal of the Western Pacific Council. The Council has been involved with native rights issues since 1986, and in 1988 it sponsored research on the historical participation of native peoples in fishing and the cultural significance of management species as part of consideration of a potential native rights preference under future limited entry management (Amesbury et al., 1989; Iversen et al., 1989; Severance and Franco, 1989). The Council also sought and obtained provisions in the Magnuson-Stevens Act that are intended to benefit the region's native fishermen and their communities. These are the Western Pacific Community Development Program or CDP, Western Pacific Community Demonstration Projects, and the Pacific Insular Area Fisheries Agreements (Sec 305 (I) and Sec 204 (e).

The Western Pacific Community Development Program provides access to fishery resources in the western Pacific for communities located in the Western Pacific Regional Fishery Management Area. These communities must meet criteria established by the Council, consist of community residents descended from aboriginal people who conducted commercial or subsistence fishing using traditional practices, and not have previously developed harvesting or processing capacity sufficient to support substantial participation in western Pacific fisheries. These communities must also develop and submit a Community Development Plan to the Council and the Secretary of Commerce detailing the process for providing access to the fishery for community residents. At this time, the Council has established and approved eligibility criteria and submitted them to the Secretary of Commerce for approval.

The Magnuson-Stevens Act also authorizes the use of grants for Western Pacific Community Demonstration Projects. The Act permits grants up to $500,000 per fiscal year to be given to not less than three, but not more than five fishery demonstration projects promoting traditional indigenous fishing practices. The Act provides for the establishment of an eight member advisory panel to review applications for grants and provide recommended rankings to the Secretary of Commerce and the Secretary of the Interior. The Council has finalized and approved the membership of the advisory panel. However, neither the Secretary of the Interior nor the Secretary of Commerce have provided funds for the 1998 fiscal year.

Finally, the Magnuson-Stevens Act provides for the establishment of Pacific Insular Area Fishing Agreements (PIAFAs). PIAFAs allow the Secretary of State in Consultation with the Council and regional elected officials to authorize foreign fishing in the Western Pacific EEZ's of American Samoa, Guam, and CNMI and to charge fees for the establishment of a Western Pacific Sustainable Fisheries Fund to be used for managing the program, and for conservation and management objectives in the western Pacific. These new provisions will allow fees and fines collected under this program to be used to develop three-year marine conservation plans that include fishery observer programs, marine and fisheries research, conservation, education and enforcement, grants to universities for technical assistance, and funding for Western Pacific Community Demonstration Projects. As of this time, negotiations with foreign nations to establish a PIAFA have not taken place, however, the most of the islands in the western Pacific have completed marine conservation plans that are under review.

There are some lessons to be learned from the Alaska CDQ experience that may help the Western Pacific Council develop and design its own programs to benefit native fishermen of the region. The Council is moving forward with planning for its demonstration projects and is considering a set-aside of permits for Native fishermen in the Mau Zone of the Northwest Hawaiian Islands Bottomfish fishery through a limited entry amendment.

BACKGROUND

The communities that may access the federal waters managed by the Western Pacific Regional Fishery Management Council include significant numbers of descendants of the aboriginal peoples of the region. The western Pacific region covers an ocean area of over 1.5 million square miles and the Council manages valuable fisheries in the EEZ while working cooperatively with international organizations (WPRFMC, 1997a; and Boehlert, 1993). The Western Pacific Regional Fishery Management Council manages 48 percent of the total U.S. EEZ. Three of the top ten ports in terms of dollar value of landed catch by domestic fishermen and by domestic and foreign fishermen in the territories of Guam and American Samoa are in the Council region (WPRFMC, 1997b). The western Pacific region has generally small island land masses and is made up of American Samoa, Commonwealth of the Northern Mariana Islands (CNMI), Guam, Hawaii, the uninhabited possessions of the United States and their respective EEZs. These U.S. associated islands and Hawaii vary significantly in land area, population, and size of their associated EEZs (Figure 6.1). Their peoples have had significantly different historical experiences, courses of political and economic development, and political relationships with the United States.

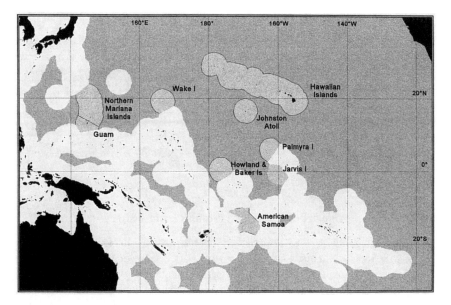

FIGURE 6.1 200-mile exclusive economic zones (EEZs) of Pacific Islands. Western Pacific Regional Fisheries Management Council EEZ areas are shown in dark gray.

Their fisheries also vary in type, size, and economic value (Tables 6.1 and 6.2). As island communities they have depended on the offshore fisheries for subsistence and ceremonial purposes for centuries. Fishery resources are thus central to the cultures of the region and to their stability and continuity. The insular nature of the communities and the cultural values of the species is recognized in the amended Magnuson-Stevens Fishery Conservation and Management Act, where it states that Pacific Insular Areas "contain unique historical, cultural, legal, political, and geographic circumstances which make fisheries resources important in sustaining their growth" (Sec. 2(10)).

TABLE 6.1 1996 Estimated Commercial Landings and Ex–vessel Value of Pelagic Species in the Western Pacific Region.

Region	Pounds	Value ($)
American Samoa	472,672	620,473
Guam	229,431	357,244
Hawaii	30,110,000	56,910,000
Commonwealth Northern Mariana Islands	224,963	431,561
Total	31,037,066	58,319,278

SOURCE: WPRFMC, 1997b

TABLE 6.2   1996 Estimated Commercial Landings and Ex–vessel Value of
Bottomfish Species in the Western Pacific Region.

Region	Pounds	Value ($)
American Samoa	32,245	62,878
Guam	54,122	17,492[a]
Hawaii	903,000	2,577,000
Commonwealth Northern Mariana Islands	52,967	176,707
Total	1,042,334	2,834,077

SOURCE: WPRFMC, 1997a
[a]Revenue based on commercial landing of 6,578 pounds

American Samoa: Community and Fishery

American Samoa is an unincorporated territory of the United States and American Samoan residents travel to and work freely in the United States. American Samoans elect their own governor, legislature (or Fono), and a non-voting representative to the U.S. Congress. The territory depends heavily on U.S. federal programs and local employment opportunities are primarily in the two large tuna canneries, government, and a newly developing garment industry. The dollar value of the catch offloaded from U.S. purse seiners, albacore trollers, and foreign longliners primarily fishing outside U.S. waters averages $200 million per year and usually places the port of Pago Pago in the top two or three ports nationwide (Hamnett and Pintz, 1997; WPRFMC, 1997a). The workforce in the shoreside canneries is almost entirely Samoan.

There is a tremendous population circulation between American Samoa and Hawaii and the continental U.S. and remittances play an important role in sustaining the Samoan community. Local legislation is protective of the culture and land rights and Samoans are in a very clear majority.   Fa'a Samoa, "The Samoan Way," or Samoan culture and lifestyle, is central to Samoan cultural identity and pride and practiced in many forms. Offshore pelagic fish, bottomfish, and nearshore reef fish play a central role in ceremonial life and subsistence. Fresh fish is expected for family and community ceremonies, title investitures, and Sunday afternoon gatherings at which the titled men are served food by the untitled men of the village (Severance and Franco, 1989).

Samoan fishermen display the full range of subsistence, ceremonial, recreational, and full-time commercial fishing (Craig et al., 1993). The decision to fish is often stimulated by cultural and ceremonial demands. Continuing access to fish is important for the maintenance of Samoan community and lifestyle. Traditionally, Samoans fished far offshore, often out of the sight of land in specialized canoes developed for trolling for skipjack tuna and other pelagic species. There is strong continuity in fishing for offshore species and bottomfish and in the use of these species for ceremonial purposes (Severance & Franco, 1989).

The Samoan commercial fishing fleet consists mostly of 28-34 ft aluminum catamarans called "Alia." These small vessels are generally powered by single outboard engines and are equipped with hand-crank longline reels, floaters, coolers, and buckets containing hooks and leaders. They are inexpensive, multipurpose vessels used for trolling for small to mid-sized pelagic fish and bottom-fishing, but they have limited capacity. (Photo by Craig Severance.)

Recent fisheries development efforts have concentrated on vessel and gear development and on small-scale shoreside ice provision. A significant portion of the local catch enters traditional avenues of distribution, but some enters the local restaurant market. This market is also serviced by direct sales of frozen fish from the canneries and from "leakage" from foreign and U.S. vessels delivering to the canneries (Kingsolving, 1996). Some foreign Asian longline caught tuna and wahoo is sold unofficially or traded to small boats in the harbor at night. To some extent, cannery sales and leakage compete with locally caught troll and longline fish in the local market and fishermen have expressed an interest in improving local and export marketing and shoreside support.

While some midsize vessels are in use, including two newly arrived longliners over 50 ft., the active Samoan commercial fishing fleet consists mostly of 28-34 ft. aluminum catamarans called "Alia" that are powered by single outboard engines. These multipurpose vessels carry coolers and wooden hand crank reels for trolling for small to mid-sized pelagic fish and bottomfishing. These vessels are locally available and relatively inexpensive to purchase and operate, but they have limited capacity. Some Alia fishermen will freeze catches of skipjack tuna until they have a full freezer load to deliver to the canneries, and these vessels tend to operate in a relatively full time commercial mode, although even these owners regularly give fish to family and friends for ceremonial use.

In 1995-96, a few Alia fishermen rediscovered harvestable quantities of albacore that could be caught by very small surface longlining gear deployed from Alias close enough to the port of Pago Pago to be delivered fresh to the canneries. The initial albacore highliners in the Alia fleet made good profits and the longline effort has been increasing as new Alias and some larger vessels enter the fleet.

Concerns have been expressed by Samoan fishermen about the possibility of U.S. registered vessels basing themselves in Pago Pago and competing with local fishermen. These concerns were stimulated in part by the rapid expansion of the Hawaii based longline fleet that led to overcapitalization and the implementation of a limited entry program in that fishery in 1994. As a result of these concerns, American Samoan fishermen and the Department of Marine and Wildlife Resources requested that the Western Pacific Council develop a limited entry amendment to the Western Pacific Pelagic Fishery Management Plan.

In August of 1998, the Western Pacific Regional Fishery Management Council voted to approve an amendment to the Pelagic Fisheries Management Plan that will create a 50 nautical mile closed area for U.S. fishing vessels larger than 50 ft. Vessels larger than 50 ft with longline permits for American Samoan waters prior to the control date would be grandfathered into the program. This amendment still needs approval by the Secretary of Commerce to become effective. The intent is to protect the developing small-scale American Samoa longline fishery from larger vessels.

At the same time, the American Samoa government has expressed interest in using the Pacific Insular Area Fishery Agreements provision of the Magnuson-Stevens Act to develop a conservation plan that will allow the licensing of foreign fishermen in the American Samoa EEZ with the revenue going to the American Samoa government to be used for fisheries development. The proposed amendment would create a small-vessel restricted zone nearer the main Samoan Islands and an unrestricted zone farther away. The proposed zone would cover approximately 80 percent of the area of the American Samoan EEZ. The intent is to protect the developing Alia based longline fishery conducted almost entirely by native fishermen while allowing the capture of some economic value from the more distant portion of the American Samoa EEZ. Management of this newly developing fishery is complicated by the fact that the nearest Pacific neighbor to the West, Samoa (formerly Western Samoa), has a large and expanding Alia longline fishery targeting albacore and delivering some of its catch to the Pago Pago canneries. The extent of the albacore resource and the possible effects of oceanographic conditions, including El Niño events, on its distribution and abundance are not well known. International cooperative management may become necessary for sustaining this resource. The implementation of Hazard Analysis and Critical Control Point regulations may create additional uncertainties in the marketing and distribution of products from this fishery (Dalzell and Schug, 1998).

Additional uncertainties in the tuna resource relate to the continued long-term viability of the canneries which have enjoyed generous tax advantages and infrastructural support from the American Samoa Government (Hamnett and Pintz, 1996). The world tuna market is highly competitive and the potential or threat of cannery closures has influenced the cannery company's relationships with the American Samoa government and the canneries' response to wage increase demands (Hamnett and Pintz, 1996). Government efforts to establish minimum wage requirements in the canneries has met with resistance. Whether support for local fisheries development projects could be obtained through negotiations with the canneries is a complex political question.

Almost all of the commercial landings by the local fleet in American Samoa come through Pago Pago since there are only three other villages on the main island of Tutuila where Alias can be berthed. The three eastern-most islands of the Manu'a group also have active Alia fishermen in the troll and bottom–fish fishery but air freight is prohibitively expensive and virtually all of the catch is used locally. Manu'a has the least developed commercial fishery of the territory and Manu'ans have had less opportunity to participate in commercial fishery development. They could be treated as a geographically separate Samoan community for CDP purposes. American Samoa also could be treated as a community as a whole (for the purposes of establishing a CDP) since economic barriers to participation on a large scale, such as a lack of capital appear to exist for most of the Samoan fishermen.

Commonwealth of Northern Mariana Islands: Communities and Fisheries

The Commonwealth of the Northern Mariana Islands (CNMI) has undergone drastic economic transformation since the commonwealth agreement came into effect in 1976 (McPhetres, 1992). The Commonwealth government maintained observer status with the Western Pacific Regional Fishery Management Council until 1996 because of the desire to control their own EEZ and the lack of a territorial sea that was under their own governance and jurisdiction. Because of the commonwealth status of the Northern Marina, federal and Western Pacific Council jurisdiction extends from the Commonwealth shoreline to 200 miles offshore. This is a politically sensitive issue. The government of the CNMI has recently agreed to continue to disagree with the United States and NMFS position on jurisdiction and to participate fully in the council process.

The CNMI Department of Fish and Wildlife (DFW) has developed a conservation plan in preparation for the possible development and implementation of

Pacific Insular Area Fishery Agreements and as a means to assess the fishery resources of their EEZ. The Commonwealth has been approached by representatives of Japan to see if the licensing of pole and line vessels to fish in the northern part of the archipelago is feasible. Little is known about the fishery resources and stock abundance, especially in the north, and there are concerns that the developing bottomfish fishery for deep- and shallow-water snappers and groupers in the northern part of the archipelago needs to be monitored.

People of the Commonwealth of the Northern Mariana Islands

The aboriginal people of the CNMI include the indigenous Chamorro as original inhabitants of the islands, and the Carolinians, who are Micronesians that resettled Saipan during the 1800s. The Chamorro represent a political majority and control much of the commercial and political activity on the islands. The Chamorros and Carolinians of CNMI elect their own governor, legislature, and non-voting representative to Congress. The Chamorro are the dominant population on the other two inhabited islands in the commonwealth: Tinian and Rota. Carolinians represent a small minority in the population but they are widely known for their seafaring and fishing skill. Much of the current population are non-resident workers from the Philippines and other parts of Asia. There are also workers from Belau and the Federated States of Micronesia, who are considered U.S. residents under their compacts of free association with the United States.

The 1995 Northern Mariana labor force had a 7 percent unemployment rate. Non-resident workers predominate the garment industry and the hotel and retail trade sectors. The Northern Mariana government provides approximately 12 percent of the jobs and the government workforce is predominantly Chamorro (Hamnett et al., in press). Other than fishing, the alternative forms of employment are generally lower wage positions and less satisfying in terms of lifestyle. Reported fishing revenues make up only a very small portion of the overall economy which has expanded dramatically in the last decade (Hamnett et al., in press). Fresh fish and fishing have important cultural significance, however, and there is an active small boat fleet that targets tunas, other small pelagics, and bottomfish. These fish are marketed locally, given away to family and friends, or used for ceremonial purposes such as parties, culturally significant fiestas, and each village's patron saint's day.

Fisheries of the Commonwealth of the Northern Mariana Islands

There is high demand for fresh and frozen product in the Commonwealth because of the substantial tourist trade and the presence of Filipino and Asian labor in the garment industry, and commercial and domestic sectors of the economy (DFW, CNMI, 1995). The local fishery is primarily a small boat troll fishery for skipjack and smaller yellowfin tuna and mahimahi to meet local market demand for fresh fish. Quality control has been a concern because few of the small-scale trolling vessels carry enough ice to adequately chill medium to large catches. The small vessels used in this fishery are often met at one of the four ramps on Saipan by fish buyers so lower quality fish are not wasted. Some skipjack is frozen by smaller local retail stores and some is marketed by roadside vendors. Unpredictability of supply and quality considerations have led much of the restaurant and hotel market to rely on primarily on imported fish.

Nearshore and bottomfish are also in high demand and there is a bottomfish export market developing for certain species. Four midsize vessels are now active in bottomfishing in the northern part of the Marina chain and the Council has strongly encouraged the CNMI DFW to conduct stock assessments of both shallow and deep water bottomfish complexes while the stocks are still underexploited. These stock assessments would provide valuable baseline data for fishery managers. Bottomfishing is generally unproductive near the main island of Saipan but there appears to be less pressure on the stocks around the islands of Tinian and Rota and the northern islands and extensive associated banks are considered underexploited.

The charterboat fishery on Saipan targets marlin and smaller pelagic fish and most of the vessels are locally owned. While most charters stick closely to Saipan and Tinian, there are some concerns about the potential impact of longer charters to the northern islands. There is also a small scale, party-boat, shallow water reef fish sportsfishery for tourists.

Tuna transshipment has been an important activity in the Commonwealth because of the closeness to Japanese markets, regular air freight service, and low costs. The large tuna storage and transshipment base on Tinian has been inactive for the last 3 to 4 years because of the bankruptcy of one large U.S. purse seiner company, a shift in purse seiner effort, and the Federated states of Micronesia's recent policy decision to require vessels licensed to fish in its waters to transship from its ports (Hamnett and Pintz, 1996). The transshipment facility continues to provide fuel and crew change service for a few longline vessels of Asian registry. The Commonwealth is considering proposals for expanding the Tinian airfield to support a reactivated transshipment base and the developing Casinos on the island.

The northern islands of the Commonwealth are largely uninhabited, but include two that have had very small resident communities. These islands are difficult to supply and lack harbors and airfields. Tinian and Rota both have small vessel troll fleets that service the local market primarily and both are planning further tourism development. Saipan itself has a well developed infrastructure and a economically stratified population. The Chamorros and Carolinians are scattered through Saipan so there is no Chamorro community in a geographic sense, unless one chooses to consider Saipan or the Commonwealth as a whole as a "community" for CDP purposes. Tinian and Rota have less developed local markets and fisheries and could be considered "communities" in a geographic sense for CDP purposes.

Guam: Community and Fishery

Guam is an unincorporated territory of the United States that is considering negotiating for Commonwealth status similar to that of the CNMI. The people of Guam elect their own governor, legislature, and non-voting representative to Congress. There is an active Chamorro (Chamoru to younger Guamanians) cultural renaissance and attempt to revitalize the language and culture (Mayo, 1992). Significant amounts of land remain in U.S. military and federal hands and the economy depends heavily on military spending and federal assistance along with a rapidly expanding tourism sector focused on the Asian tourist market.

People of Guam

Chamorro's represent approximately 40 percent of the total resident population, which also includes military personnel and their dependents, alien workers, recent migrants from the Federated States of Micronesia, Palau, the Marshall Islands, the Northern Mariana Islands, and residents from Hawaii and the U.S. mainland. Fish are important for subsistence and for ceremonial use at parties, annual village fiestas, and other family events and there is heavy pressure on local reef fish and bottomfish stocks. Locally caught pelagic fish are also in demand, although there is not enough supply to meet demand.

Guam is a single island with the smallest EEZ in the region and it has relatively few offshore banks. The local small boat fishery is primarily oriented to seasonal trolling for pelagic fish (Myers, 1993; Hensley & Sherwood, 1993). There has been limited and economically unsuccessful experimentation with small scale longlining. Guam has a well developed port and airfreight infrastructure and its proximity to Asian markets has led it to become an important tuna transshipment and vessel servicing facility. Foreign and domestic purse seiners and some foreign longliners transship, refuel, reprovision, exchange crews and provide crew rest and relaxation in Guam. This is a significant contribution to the Guam economy ranking Agana, Guam, the fourth largest port in the United States, in overall dollar value of commercially landed catch, at $91 million in 1996 (Table 6.3). Local revenues generated by transshipment activities are limited and difficult to assess (Hamnett & Pintz, 1996).

In the last two years there has been a reduction in the Guam based U. S. purse seine fleet due to the same reasons noted above for the Commonwealth. Some lower quality tuna enters the local market through the transshipment process and this is perceived by some fishermen to compete with them in the local market (Hamnett & Pintz, 1996). The government of Guam is investing heavily in expanding port facilities in hopes of maintaining its importance as a transshipment base.

Guam has also invested in marina expansion to support the active charterboat fleet and tourism development. It appears that much of the fish caught by the

TABLE 6.3   1996 Exvessel Value of Commercial Fishery Landings by
Domestic and Foreign Vessels at Major U.S. Ports.

Port	Value of Landings (Millions of Dollars)
Pago Pago, American Samoa	211.8
Dutch Harbor-Unalaska, Alaska	118.7
New Bedford, Massachusetts	100.5
Agana, Guam	94.2
Kodiak, Alaska	82.3
Key West, Florida	62.8
Brownsville-Port Isabel, Texas	60.0
Honolulu, Hawaii	50.1
Point Judith, Rhode Island	46.0
Empire-Venice, Louisiana	45.4

SOURCE: WPRFMC, 1997a

small boat fleet is distributed through personal and family ties as well as through
roadside vending. As in the Commonwealth, sashimi grade tuna for the hotel and
restaurant market tends to come through reliable and predictable import sources
(Bartram et al., 1996). The size of the trailerable small boat fleet, current levels
of effort, the relative mix of commercial and recreational fishermen, and the ethnic make up of the participants is not well known, although research on the social
makeup and cultural significance of the Guam fishery is in progress.

The Chamorro population is scattered throughout Guam so a geographically
defined community is difficult to determine for the purposes of establishing a
CDP, except by defining the Chamorro population on Guam as a single community.

Hawaii: Communities and Fisheries

Hawaii is a densely populated state that includes the inhabited Main Hawaiian Islands (MHI) and largely uninhabited (except for military support and research personnel) Northwest Hawaiian Islands (NWHI). Hawaiians are 20 percent of the total population of over one million. They are the most rapidly growing
portion of the population and by most statistical measures they have the lowest
incomes and poorest health care of any ethnic group in the state. Federal, state,
and private programs exist to benefit Hawaiians but there is a widespread perception that these are not adequate or always well managed. There is an active
cultural renaissance with efforts to restore the language, the arts, and subsistence
activities, including traditional fishing practices. There is an active effort to gain
control of ceded lands and their revenues by some sovereignty groups. Gathering rights and nearshore fishery rights are an active area of political concern and
interest in offshore fishery rights is developing in the Hawaiian community.

Recently, new efforts have been made by the State of Hawaii to strengthen native Hawaiian fishing heritage on the island of Kaho'o'lawe which had been used as a naval bombing site until 1988. The removal of munitions began in 1993 and a Kaho'o'lawe Island Reserve Commission has been established to promote traditional Hawaiian culture and traditional fishery practices on the island. There is also a community management demonstration project underway at Mo'omomi on Molokai that regulates fishing gear and access to local fisheries through community management. In some regions of Molokai, and other islands, there has been increased interest in the use of traditional fish ponds.

The NWHI are under joint federal (NMFS and U.S. Fish & Wildlife Service) and state jurisdiction and the archipelago includes a number of islands that serve as a wildlife refuge for threatened and endangered sea birds, marine turtles and monk seals. Exclusion zones for longliners, bottomfishermen and lobster-fishermen limit interactions with these species and land access is generally only by permit.

Because of its status as a state, the fisheries interests in Hawaii are comparatively better represented in Congress than those of the commonwealth and the two territories, although the Hawaii congressional delegation is sensitive to and supportive of the interests of the western Pacific region as whole. The Western Pacific Regional Fishery Management Council is based in Hawaii, although it holds meetings in other parts of the region on a regular basis and it seeks to operate by consensus in a Pacific Island style. Given the limited size of the continental shelves around the islands, the pelagic nature of much of the fishing activity, and apparently healthy fish stocks, the Council has not had to deal with as many contentious issues as the North Pacific Fishery Management Council.

The Western Pacific Council has taken a proactive management role by establishing limited entry and a quota (and a TAC) in the Northwest Hawaiian Lobster fishery. The Hawaii based longline fishery has a set number of permits but no quota, and a portion of the Northwest Hawaiian Islands bottomfish fishery also has a set number of permits but no quota. The NWHI bottomfish fishery is currently divided into two zones: a distant limited entry zone, with a small number of permits, and an open access zone nearer to the Main Hawaiian Islands where people with permits may develop fishing history and potentially qualify for permits to fish in the limited entry area. Stocks in the more distant zone appear to be healthy based on annual catch per unit effort data but the number of new permits is not likely to increase. Concerns about pressure on the nearer zone stocks have led to development of a Western Pacific Council task force and a draft amendment to the bottomfish fishery management plan to make this zone limited entry also. The Council approved an amendment to the bottomfish Fishery Managment Plan (FMP) to make the Mau Zone a limited entry zone. Two permits (20 percent of the target number of permits) have been set aside for Native Hawaiian fishermen under a CDP designation. This amendment is under review by the NMFS Regional Director and will require Secretary of Commerce approval to take effect.

Northwest Hawaiian Islands Lobster Fishery

The Northwest Hawaiian Islands lobster fishery has a small number of permitted vessels, including some that come from the U.S. West coast to participate. The annual quota is set based on surveys, stock assessment models, and the commercial catch data from the previous year. The lobster stocks are affected by climatic events (Polovina et al., 1994) and this fishery suffered an apparent stock collapse and emergency closure in 1995. The fishery is rebuilding slowly and appears to be profitable for the small number of participants, although uncertainty continues. There are more qualifying vessels than those which have fished over the last two seasons so opportunities for new permits are limited. Fractional licensing or permit sharing for this fishery may be a management option (Townsend and Pooley, 1995). Should this fishery stabilize and expand, sources of capital might allow an increase in participation by Hawaiians through a small CDP.

Hawaii Longline Fishery

The Hawaii-based longline fishery was placed under limited entry in 1995 after rapid expansion of the fleet and the arrival of many U.S. flagged vessels from outside Hawaii. Perceived competition with the small scale troll and handline fleet, and some conflicts over the fishing gear used, led to the creation of longline exclusion areas around the Main Hawaiian Islands. A permit moratorium was followed by the limited entry program created by Amendment 7 to the Pelagics Fishery Management Plan. Currently, there are fewer vessels actively fishing than permitted and there is economic uncertainty depending on vessel size and targeting strategy (Hamilton et al. 1996). However, the value of the longline catch is significant enough to place Honolulu in the top ten ports for fish landings in the United States. This fishery is currently managed not by a quota, but by limiting permits to vessels and vessel owners and by making permits transferable. A set-aside of permits through transferability measures may be a means by which additional Hawaiian participation could be encouraged in this fishery.

Hawaii Bottomfish Fishery

Bottomfish stocks in the Main Hawaiian Islands are seriously overfished. Since approximately 80 percent of the bottomfish habitat in the MHI lies within state waters, the Council has given responsibility for MHI bottomfish management to the State of Hawaii but has maintained oversight and has encouraged the state to take action to rebuild the stocks. The Hawaii Department of Land and Natural Resources approved regulations to establish closed areas for bottomfishing and may consider limited entry in the future. The Council is supportive of the state's efforts but has also developed a backup plan to close federal waters in the Main Hawaiian Islands to bottomfishing if necessary. Overfishing provisions in the Magnuson-Stevens Act require the Secretary of Commerce to take action to rebuild stocks if the state's plan is deemed to be ineffective or the Council does not act.

Recreational and Charter Fishing

Hawaii also supports an active commercial troll and handline fishery for pelagic fish and a large charterboat fishery (Walker, 1996). Charter boats are commercially licensed and the fish may be sold. The size of the active recreational troll fishery is not known, although one small boat survey conducted prior to the expansion of the longline fleet suggested that the recreational catch may be equal to the commercial sector in terms of poundage for pelagic fish (Meyer, 1987). A subsequent small boat survey also indicated that the recreational catch which is not regularly reported in Hawaii is a significant portion of the overall catch (Hamm and Lum, 1992).

Hawaiians participate in this fishery and these fish have important cultural significance. Hawaiian participation appears to range from fishing primarily for subsistence and customary exchange with family and friends to full-time commercial fishing. Possible historic and economic barriers to Hawaiian participation in these fisheries, and especially to the larger scale longline fishery, are not well known but appear to include lack of capital and of training in business skills.

There are some small communities made up predominantly of Hawaiians with various degrees of Hawaiian ancestry. These communities include Ni'ihau, Milolii, and Wai'anae and the island of Molokai. However, much of the Hawaiian population is scattered throughout the main Hawaiian Islands so that a geographic community definition for CDP purposes may be difficult unless one chooses to define the Hawaiian community as all residents of the state with Hawaiian ancestry. Most federal and state programs define eligibility for Hawaiian native programs as those individuals who can trace their ancestry to persons resident in Hawaii prior to the arrival of Captain James Cook in 1779.

Hawaii Precious Coral Fishery

The Western Pacific Regional Fishery Management Council also manages precious corals through a Fishery Management Plan with an adjustable quota. There has been no harvesting since the FMP was created although one company has received a federal permit for an experimental fishery. This is a capital-intensive high technology fishery requiring the use of submersibles or remotely operated vehicles and its potential value is unclear. The Council is also in the process of planning for development of a coral reef fishery management plan that would use an ecosystem approach to manage coral reef resources including invertebrates, corals, and finfish of commercial value.

Uninhabited Possessions

The uninhabited island possessions of the United States including Baker Island, Howland Island, Jarvis Island, Johnstol Atoll, Kingman Reef, and Palmyra Atoll are also subject to the jurisdiction of the Western Pacific Council. These regions are relevant to future western Pacific community development programs because provisions for Pacific Insular Area Fisheries Agreements under the Magnuson-Stevens Act will allow access fees, or fines levied on foreign vessels caught poaching in the EEZ surrounding these possessions to be used for a variety of conservation and management plans. Part of these funds could be used for community development fishery projects. Although harvests in the EEZ surrounding these uninhabited possessions is not well documented, improved monitoring and enforcement under PIAFAs could provide revenue for a variety of projects in the western Pacific region.

FINDINGS

1. Community Development Quota Programs have not been created for the fisheries of the western Pacific Region so their effectiveness cannot be evaluated in terms of objectives or benefits.

2. Direct transfer of the Alaska CDQ model and its specific provisions is inappropriate for the western Pacific Region because of significant differences between these regions in fisheries conditions, fishery management strategies, and the nature of the native communities.

3. The purposes of CDQ programs may be supported by alternative community development programs developed by the Western Pacific Council. Sources of capital other than quota shares could achieve these purposes. Potential sources of capital include foreign fishing fees through Pacific Insular Area Fishing Agreements and fines levied against foreign vessels caught fishing in the U.S. EEZ.

4. The Western Pacific Council is developing plans for encouraging native communities to create Community Development Plans that are intended to increase participation by Native fishermen in the fisheries of the region, and has developed criteria for participation. Such programs have similar purposes to CDQs and like CDQs need to be tailored to the specific conditions of the region. No formal Community Development Plans have created as of this time, however, the Council has set aside a number of limited access permits for Native Hawaiians under a future CDP for the Mau Zone bottomfish fishery. Fishery workshops have been conducted for American Samoa and the Commonwealth of the Northern Mariana Islands to further discuss the concept of a CDP in these regions.

5. Only the small and highly variable Northwest Hawaiian lobster fishery and the inactive precious coral fishery are managed by quota in the western Pacific Region. These two fisheries are not currently large or valuable enough to extract significant economic benefit or capital that could be allocated to Native fishermen through a CDQ-like program.

6. All other fisheries in the western Pacific Region are managed as open access or limited entry fisheries. Those that have limited entry could be adjusted to encourage more participation by native fishermen through permit transferability clauses, permit set-asides, or new sources of capital through community development programs. Alternative measures might be used to encourage participation by native fishermen, such as modifications to gear restrictions, fishing days, or special fishing zones.

7. Sectors of western Pacific communities display social conditions similar to those in the Western Alaska CDQ communities, including high unemployment, a lack of infrastructure for fisheries, and a variety of social ills.

8. The specific criteria used for defining eligible communities in Alaska may be inappropriate for the western Pacific. For instance, geographically based criteria would be difficult to apply in some places in the western Pacific because communities are widely dispersed, although it might be possible to treat whole archipelagos as one community, as in American Samoa or use geographic criteria in certain isolated locations such as the islands composing the Manu'a group and Swain's Island in American Samoa, Tinian and Rota in the Northern Mariana Islands, and possibly certain communities in the Hawaiian Islands.

9. Some of the elements of the CDQ program in Alaska may be relevant and instructive to the western Pacific. If CDQs are pursued in the western Pacific, attention should be given to how Alaska addressed administrative structure, costs, oversight of investment strategies, selection of personnel, business training, and performance evaluation.

FINAL THOUGHTS

Lessons learned from the Alaskan experience with Community Development Quotas can help in the design of similar programs elsewhere. The Western Pacific Regional Fishery Management Council should take into consideration the Alaskan experience with Community Development Quota fisheries if it plans Community Development Programs in its region and tailors them to the specific conditions of western Pacific fisheries and communities. Steps could be taken to increase the fishing opportunities and degree of participation by native fishermen in western Pacific communities that borrow generally from the Alaskan experience and are tailored by the Western Pacific Council to the conditions of the region without trying to duplicate the CDQ program, which is designed to suit the specific conditions found in Alaska.

Western Pacific Community Development Programs would need to define realistic goals that fit within Council purposes and plans. Definitions of eligible communities would need to be crafted carefully so the potential benefits accrue in an equitable fashion to native fishermen. If Community Development Programs show promise and potential, the Council could, over time, investigate additional sources of capital and other avenues such as additional permits and permit transfers to encourage greater participation in the fisheries by native people of the region. As the Western Pacific Council considers the Alaskan CDQ experience and the differential performance of CDQ groups, it should recognize that CDQs constitute only one possible model for community development in fisheries.

7

Conclusions and Recommendations

The Community Development Quota (CDQ) program was implemented in December 1992 by the North Pacific Fishery Management Council (NPFMC). It was designed to be an innovative attempt to accomplish community development in rural coastal communities in western Alaska, and in many ways it appears to be succeeding. The CDQ program has fostered greater involvement of the residents of western Alaska in the fishing industry and has brought both economic and social benefits. The program is not without its problems, but most can be attributed to the newness of the program and the inexperience of participants. Overall, this committee finds that the program appears on-route to accomplishing the goals set out in the authorizing legislation: to provide communities with the means to develop ongoing commercial fishing activities, create employment opportunities, attract capital, develop infrastructure, and generally promote positive social and economic conditions.

Because the program is still relatively new, the data necessary for detailed evaluation are limited and it is not yet possible to recognize long-term trends. Generally, however, for a young program this committee sees promise. The six CDQ groups, organized from the 56 eligible communities (later expanded to 57), were of various sizes and took different approaches to harvesting their quota and allocating the returns generated. Although not all groups have been equally successful, there were significant examples of success. All six groups saw creation of jobs as an important goal, and stressed employment of local residents on the catcher-processor ships. All incorporated some kind of education and training component for residents, although to different degrees and with different emphases. The CDQ program has one very evident strength: it gives local communities increased control of their own destinies. The program is also well-designed in that the fisheries base gives options for the local people to continue some elements of their subsistence lifestyles, given the periodic nature of the ship-board employment. The State of Alaska also has played its part relatively effectively—it was efficient in reviewing the Community Development Plans, monitoring how the communities progressed, and responding to problems. Some of these responses, like reallocating quota, have been controversial, as might be expected.

Perhaps the greatest weakness of the CDQ program as implemented is lack of open, consistent communication between the CDQ groups and the communities they represent, particularly a lack of mechanisms for substantial input from the communities into the governance structures. There has also been a lack of outreach by the state to the communities to help ensure that the communities are aware of the program and how to participate. Some controversy has surrounded the uncertainty about the intended beneficiaries of the program—essentially, whether the program is intended primarily for the Native Alaskan residents of the participating communities, and, if not, review the governance structures to ensure that non-native participation is possible. Similarly, there has been dissatisfaction among segments of the fishing industry that are not involved, either directly or as partners of CDQ groups, that the program unfairly targets a particular population for benefits; this conflict is inevitable, given that the CDQ program is designed to provide opportunities for economic and social growth specifically to rural western Alaska. This policy choice specifically defines those to be included and cannot help but exclude others.

To accomplish its goals, the CDQ program must, over the long-term, use the quota allocation in ways that generate funds that are then used to enhance the communities' continued participation in the fishing industry and to expand their non-fishery related activities. This committee believes that the CDQ program, while a financially modest effort at $20 million per year, shows great promise.

While this report reviews the CDQ program in a broad way, there is need for periodic, detailed review of the program over the long term (perhaps every five years). Such a review should look in detail at what each association has accomplished—the nature and extent of the benefits and how all funds were used.

The committee warns that for a program like this, care must be taken not to use strictly financial evaluations of success. Profits gained from harvest per year and numbers of local people trained are valuable measures, but they must be seen within the full context of the program. It is a program that addresses far less tangible elements of "sustainability," including a sense of place and optimism for the future.

CONCLUSIONS AND RECOMMENDATIONS

The committee offers the following main conclusions and recommendations, which are discussed in more detail in Chapters 4, 5, and 6:

Community Development Strategies

Although the Community Development Plans developed by the different CDQ groups are similar in some important respects, the specific elements included vary considerably. Each CDQ group derives income from the large scale pollock fishery through royalties and employment, and each seeks to develop nearshore fisheries using smaller vessels. The diversity of infrastructural investments, training programs, and financial strategies adopted by the CDQ groups does, in our judgment, appropriately reflect varying circumstances and reasoned approaches to diverse problems. To some extent the development plans were shaped by uncertainty about the duration of the CDQ program and by the restriction that the CDQ plans must focus on fishery development. For example, the uncertainty may have encouraged at least one CDQ group to seek a quick financial gain through sale of their processing quota rights in perpetuity. We found this permanent conveyance to be inconsistent with the philosophy and intent of the CDQ program. Finally, the economic and cultural development of these communities may at times be advanced through non-fishery employment or investments. Hence, we found no strong reason to require the communities to use funds generated from their CDQs to invest only in fisheries.

Recommendations

• We recommend that the State of Alaska prohibit permanent conveyance of community development quotas into the hands of commercial enterprises outside the communities. An important aspect of the community development sought in western Alaska is the continuing and direct involvement of local people in fisheries of the Bering Sea. Sale of the CDQs to commercial interests outside the communities will create an inappropriate separation of the people from the regional resources.

• We recommend that the restriction that CDQ revenues to be invested only in fishery-related activities should be removed, at least for some portion of the revenues. Many of the communities will find that fishery investments are still the ones they wish to undertake. However, since community development is broader than fishery development, funds should also be available for other activities that will enhance community infrastructure or land-based economic activity. This broadening of the allowed investments would also remove uncertainty about whether particular investments are indeed "fishery related" and thus allowable under current rules.

Participation and Benefits

The CDQ program has had an important positive economic impact on western Alaska communities. Significant revenues have been generated and

employment has been enhanced, especially for the mobile members of the community. In addition, the general educational and training programs have been as beneficial as specific fisheries employment.

Recommendations

- The CDPs should be careful to balance the mix of local fishing with wage-earning opportunities with fishing partners. This is important because local fishery development can occupy less mobile village residents, while wage-earning opportunities in the industrial fleet are especially important for younger adults. A focus on local fisheries opportunities, where they exist, for permanent village residents will more closely tie the CDQ program to the village economies.

- To improve the effectiveness of developing a well trained workforce, the CDQ groups need a strategic plan for education and training programs. This would include internships and technical training for direct employment with the industrial fishing partners of the CDQ groups, formal university education in fields pertinent to the development goals of native residents, and training of administrators and board members of CDQ organizations. The ultimate objectives would be to develop both the business acumen and labor productivity of village residents.

Governance and Decision-Making

The CDQ groups were given a unique governance structure that includes elements of both State and federal oversight, which is appropriate given the goals of the program. But the extensive and variable criteria used by the State and federal governments in allocating quota among the groups causes decisionmaking to be inconsistent and difficult to evaluate. That the lists of evaluation criteria are not entirely consistent with one another in either content or order of listing presents additional opportunity for confusion among the CDQ groups and the public in evaluating the logic and fairness of the decisions made by the governor and ratified by the Secretary of Commerce.

Recommendations

- State and federal criteria for the allocation of quota based on performance and plans should be less complicated than they are and should also be consistent with one another. We recommend that changes be made to simplify the criteria, in consultation with the CDQ groups.

- The committee notes that the criteria currently are used for two purposes: to allocate quota equitably and to encourage good management. One way to clarify some of the confusion created by using the criteria in this way would be to separate these two purposes into two allocations of quota. A "foundation quota" would address issues of equity and a "performance quota" would address issues of performance. The foundation quota (likely more than half of the allocation) would be allocated on measures of population, income, employment, and proximity to the fishery being allocated. The performance quota (the remainder) would be allocated based on clearly defined performance measures such as accomplishments of the CDP goals, compliance with fishing regulations (e.g., regarding bycatch), quality of community development plans, and so forth.

• One way to improve responsiveness of the CDQ groups' managers to the communities would be to improve communication. Although the idea of locating the headquarters of the CDQ groups near potential business partners and the State government may have made sense in the early years of the program, as it matures and the management proves its business capability, relocation of the headquarters to the communities may have significant benefits in terms of responsiveness to the desires of the community members.

• Communication would be further improved if the confidentiality rules and the rules for making information available to constituents were improved. NMFS and the state needs to collaborate to resolve any potential conflicts between state laws regarding the confidentiality of financial data and the evaluation of the CDQ program objectives. Information on the number of people employed by the program and the earnings in each of the communities should be provided.

• Although some of the CDQ groups have created newsletters, a requirement that newsletters to communicate with constituents, town meetings, or other forms of communication appropriate to reach community members might be a helpful step in improving communication in the communities.

Development of Human Resources

Education, training, and other activities to develop human resources in the participating communities are an explicit part of the CDQ program mandate and a key element in ensuring the program's success because stable, healthy communities depend as much on people as on economics.

Recommendations

• To be truly effective, the CDQ groups must have education and training elements. These elements should not be haphazard, but carefully planned and coordinated so they meet community needs. Both vocational training and support for higher education will help members of the community acquire the skills and knowledge needed for more advanced technical and managerial positions. The number of people receiving education and training should be provided.

• CDQ groups need to do a better job disseminating information that describes the educational and training opportunities open to the use of program funds. They also need to improve their recordkeeping of education and training initiatives so the results can be monitored over time. A common framework for recording and reporting their efforts would be useful.

Program Duration

The CDQ program must be a long-term program because it deals with a long-term issue: development of healthy, sustainable communities in coastal Alaska. Long-term economic development requires stability in the under-lying policy base so decision-makers can make choices that balance current and future needs.

Recommendations

• The original CDQ program was a three-year trial. It was subsequently extended and then made a more permanent part of the fishery management system with the passage of the Magnuson-Stevens Fishery Conservation Act reauthorization in 1996. This program has been successful in bolstering community development in western Alaska. It has passed a crucial point in its evolution and we should expect the allocation of harvests to the CDQ groups to become a long-standing, if not permanent, feature of the federal fishery management system in the North Pacific.

• The committee recommends that the CDQ program should be reviewed on a periodic basis to determine if the preliminary trends observed by the committee continue in the future. Reviewing the CDQ program in another five years may provide important additional information on the effects of the program and provide valuable suggestions for its management.

Economic Sustainability and Environmental Stewardship

Economic sustainability implies programs and policies that offer the greatest assurance of economic options over the long-term to a population that chooses to remain in specific locations. That is, given alternative economic futures for a people (or for a community), economic sustainability would entail choosing that future with the lowest probability of inducing economic decline as measured by a range of indicators. Economic sustainability is but one part of the larger problem of ecological and socio-cultural sustainability. Clearly, communities that squander their local environmental resources (or that fail to maintain cultural and social processes and structures) will be incapable of economic sustainability. Large-scale commercial fishing activities can have negative impacts on ecosystems, either independently or through interaction with natural fluctuations. Because the CDQ program is designed specifically to increase participation in fisheries activities and at the same time improve the long-term economic conditions of the

participating communities, greater emphasis should be given to environmental stewardship.

Recommendations

• Concern for the long-term health of the Bering Sea ecosystem needs to feature more prominently in the CDQ program. Local concerns about environmental stewardship need to be able to be expressed in a meaningful way throughout the program's management structure, beginning with effective communication of local concerns to the CDQ group management and continuing on up through the Council process. The quota allocation process can be used to increase the emphasis on environmental stewardship.

• Economic sustainability is dependent upon sound environmental stewardship. In order for the CDQ program to help build a sustainable economy in the region, it is imperative that the underlying resource base—the fisheries—be used in ways that are sustainable over the long-term. This will require explicit, in-depth, continuing analysis of the condition or health of the fishery resource and management that can respond and adapt to changes in this condition.

Relevance of the CDQ Experience to the Western Pacific

The CDQ program was designed specifically to address the issues and environment of western Alaska and thus is not appropriate, in its current form, for the Western Pacific Region. If similar goals such as inclusion of native communities in fisheries are desired in the region, a program could be tailored to the conditions of the western Pacific, although fisheries in the region are not now generally managed by quota. There should be real efforts to communicate the nature and scope of the program to the residents of the participating villages, and to bring state and NMFS managers to the villages to facilitate a two-way flow of information. In addition, geographic criteria for eligibility would be difficult to apply because the communities are widely dispersed. As the Western Pacific Regional Fishery Management Council considers the Alaskan CDQ experience and the differential performance of the CDQ groups, it should recognize that CDQs constitute only one possible model for community development in fisheries. But if CDQ-type programs are seriously considered for the western Pacific the committee recommends:

• CDQ-type programs in the western Pacific would need to define realistic goals that fit within Council purposes and plans, and definitions of eligible communities would need to be crafted carefully.

• To assist in the design of such programs, lessons can be learned from detailed study of the Alaskan experience related to program structure, costs, oversight, performance evaluation, and other administrative issues.

FINAL THOUGHTS

What emerges from a review of the western Alaska CDQ program is an appreciation that the program is as an example of a broad concept adapted to very particular circumstances. Others interested in the application of CDQ-style programs are likely to have different aspirations and different contexts. Wholesale importation of the Alaskan CDQ program to other locales is likely to be unsuccessful unless the local context and goals are similar.

Any new program, especially one with the complex goal of community development, should be expected to have a start-up period marked by some problems. During this early phase, special attention needs to be given to work out clear goals, define eligible participants and intended benefits, set appropriate duration, and establish rules for participation. In addition to these operational concerns, those involved—the residents and their representatives—must develop a long-term vision and coherent sense of purpose to guide their activities.

For the CDQ program to be effective there must be a clear, well-established governance structure that fosters exchange of information among the groups' decision-makers, the communities they represent, and the state and federal personnel involved in program oversight. Greater openness of information is critical, as is regular detailed review.

Although it is logical to require initially that all reinvestment of profits be only in fishery-related activities because the initial objective of the CDQ program is to help the participating communities establish a viable presence in this capital intensive industry, over time there should be more flexibility in the rules governing allocation of benefits—perhaps still requiring most benefits to be reinvested in fishing and fisheries-related activities but allowing some portion to go to other community development activities. This will better suit the long-term goal of the program, which is development of opportunities for communities in western Alaska.

The main goal of the CDQ program—community development—is by definition a long-term goal. Thus there is a need for a set and dependable program duration and the certainty that it brings to oversight and management. This will allow CDQ group decision-makers to develop sound business plans and reduce pressures to seek only short-term results. However, calling for the program to be long-term does not mean it must go on indefinitely nor that it must never change. Periodic reviews should be conducted, and changes made to adapt rules and procedures as necessary. There can be a balance between certainty and flexibility if the program is assured to exist for some reasonable time and if major changes in requirements are announced in advance with adequate time to phase in new approaches. The appropriate time scales will of course vary with the nature of the change, with minor changes requiring little notice and major changes requiring enough time for decision-makers and communities to plan and adjust.

References

Ackerman, R. 1964. Prehistory in the Kuskokwim-Bristol Bay Region, Southwestern Alaska. Laboratory of Anthropology, Report of Investigations-26. Pulman, Washington: Washington State University.

Ackerman, R. 1984. Prehistory of the Asian Eskimo zone. in D. Damas, ed. Arctic, Handbook of the North American Indian. Vol 5:106-118. Washington, D.C.: Smithsonian Institution Press.

Alaska Department of Community and Regional Affairs Database. 1998. http://www.comregaf.state.ak.us/CF_ComDB.htm

Alaska Department of Fish and Game (ADF&G). 1997. Annual Management Report for the Shellfish Fisheries of the Westward Region, 1995. Regional Information Report No. 4K97-16, 225pp. Alaska Department of Fish and Game, Commercial Fisheries Management and Development Division, March, 1997. Kodiak, Alaska.

Alaska Fisheries Science Center (AFSC). 1998. Commercial Fisheries Landings in Alaska for 1996. National Oceanic and Atmospheric Administration (NOAA), Alaska Fisheries Science Center (AFSC).

Alaska Natives Commission. 1994. Alaska Natives Commission, Final Report. Anchorage, Alaska.

Amesbury, J.R., Hunter-Anderson, R.L., and Wells, E.F. 1989. Native Fishing Rights and Limited Entry in the CNMI. Western Pacific Regional Fishery Management Council Report. Guam: Micronesian Archeological Research Service.

Anchorage Daily News. 1989. People in Peril (Series of Articles). Special Reprint. Sunday January 15, 1989 - Tuesday January 17, 1989. Anchorage, Alaska.

Arnold, R. 1976. Alaska Native Land Claims. Anchorage: Alaska Native Foundation.

Arrow, K.J. and Raynaud, H. 1986. *Social Choice and Multicriterion Decision-Making.* Cambridge, Massachusetts. The MIT Press.

Baduini, C.L., Hunt, Jr., G.L., and Hyrenbach, K.D. 1998. Die-off and Starvation of Short-Tailed Shearwaters *(Puffinus tenuirostris)* in Relation to Prey Availability in the Eastern Bering Sea. Abstract submitted to PSG meeting in Monterey, California. *Pacific Seabirds* 25(1):21.

Bartram, P., Garrod, P., Kaneko, J. 1996. Quality and product differentiation as price determinants in the marketing of fresh Pacific tuna and marlin. Honolulu: SOEST Publication 96-06, JIMAR Contribution 96-304.

144

Berger, T.R. 1985. Village Journey: Report of the Alaska Native Review Commission. New York: Hill and Wang.

Bockstoce, J. 1986. *Whales, Ice, and Men: The History of Whaling in the Western Arctic.* Seattle: University of Washington Press.

Boehlert, G. 1993. Fisheries and marine resources of Hawaii and the U.S.-associated Pacific Islands: an introduction. *Marine Fisheries Review* 55:3-7.

Browning, Robert J. 1980. Fisheries of the North Pacific: history, species, gear and processes. Anchorage: Alaska Northwest Publishing Co.

Case, D.S. 1984. Alaska Natives and American Laws. Fairbanks: University of Alaska Press.

Chris Oliver (North Pacific Fishery Management Council) to Eugene Smith (Chukchi Sea Fishermen's Cooperative). 11 May 1993.

Clark, D.W. 1984. "Pacific Eskimo: historical ethnolography," in Handbook of Native American Indians: Volume 5 Arctic. David Damas ed. Washington: Smithsonian Institution Press. pp. 185-197.

Cobb, J. 1927. Preliminary Report of Fishway Work. University of Washington Press. Seattle, Washington.

Commercial Fisheries Entry Commission (CFEC). 1996. Changes Under Alaska's Halibut IFQ Program, 1995. CFEC Report 96-R10N. Juneau, Alaska.

Commercial Fisheries Entry Commission (CFEC). 1998. Holdings of Limited Entry permits, Sablefish Quota Shares, and Halibut Quota Shares through 1997 and Data on Fishery Gross Earnings. CFEC Report 98-SP. Juneau, Alaska.

Cotton, Lamar (Department of Community and Regional Affairs) personal communication to Steven Pennoyer (Nationl Marine Fisheries Service). 28 January, 1998.

Craig, P., Ponwith, B., Aitaoto, F., and Hamm, D. 1993. The commercial, subsistence and recreational fisheries of American Samoa. *Marine Fisheries Review* 55:109-116.

Dalzell, P., and Schug, D. 1998. Current Status of Pelagic Fisheries of the Western Pacific. Honolulu: Western pacific Regional Fishery Management Council. unpublished manuscript.

Department of Community and Regional Development (DCRA). 1997. Multi-species Community Development Quota Program: 1998–2000. Community Development Plan Application Review. Juneau, Alaska.

Division of Fish and Wildlife, DLNR, Commonwealth of Northern Mariana Islands. 1995. Analysis of Saipan's Seafood Markets. Saipan, CNMI: DFW/DLNR.

Dumond, D. 1984. "Prehistory of the Bering Sea Region." in D. Damas, ed. Arctic, Handbook of the North American Indian. Vol 5:94-105. Washington, D.C.: Smithsonian Institution Press.

Dunlop, H.A., Bell, F.H., Myhre, R.J., Hardman, W.H., and Southward, G.M. 1964. Investigation, Utilization and Regulation of Halibut in Southeastern Bering Sea. International Pacific Halibut Commission Report 35. Seattle, Washington: International Pacific Halibut Commission.

Federal Register (FR), 53:8938, March 18, 1988.

Federal Register (FR), 53:10536, April 1, 1988

Federal Register (FR), 53:20327, June 3, 1988.

Federal Register (FR). 1992. 57. No. 248:61329.

Fienup-Riordan, A. 1983. *The Nelson Island Eskimo.* Anchorage: Alaska Pacific University Press.

Fienup-Riordan, A. 1986a. "Traditional Subsistence Activities and Systems of Exchange Among the Nelson Island Yup'ik," in Steve Langdon, ed., *Contemporary Alaskan Native Communities.* Lanham, MD: University Press of America, pp. 173-183.

Fienup-Riordan, A. 1986b. When Our Bad Season Comes: A Cultural Account of Subsistence and Harvest Disruption on the Yukon Delta. Aurora Monograph Series 1. Anchorage: Alaska Anthropological Association.

Fienup-Riordan, A. 1990. *Eskimo Essays.* New Brunswick: Rutgers University Press.

Fienup-Riordan, A. 1994. Boundaries and Passages: Rule and Ritual in Yup'ik Eskimo Oral Tradition. Norman, Oklahoma: University of Oklahoma Press.

Fitzhugh, W. and Kaplan, S. 1982. *Inua: Spirit World of the Bering Sea Eskimos.* Washington, D.C.: Smithsonian Institution Press.

Fitzhugh, W. and Crowell, A. 1988. *Crossroad of Continents: Cultures of Siberia and Alaska.* Washington, DC: Smithsonian Institution Press.

Gessener, B.D. 1997. Temporal trends and geographic patterns of teen suicide in Alaska. *Suicide and Life Threatening Behavior,* 27:60-67.

Ginter, J. 1997. Presentation to the National Research Council Committee to Review Community Development Quotas. August 6, 1998. Girdwood, Alaska.

Hamilton, M.S., Curtis, R.E., Travis, M.D. 1996. Hawaii longline vessel economics. *Marine Resource Economics.* 11:137-140.

Hamm, D.C., and Lum, H.K. 1992. Preliminary Results of the Hawaii Small Boat Fisheries Survey. Honolulu: Southwest Fisheries Science Center, NMFS/NOAA. Admin. Rep. H92-08.

Hamnett, M., Anderson, C., Franco, R., Severance, C. in press. Coordinated Sociocultural Investigation of Pelagic Fishermen in the Commonwealth of the Northern Mariana Islands. SOEST Publication. JIMAR Contribution forthcoming.

Hamnett, M., and Pintz, W. 1996. The contribution of tuna fishing and transshipment to the economies of American Samoa, the Commonwealth of the Northern Mariana Islands, and Guam. Honolulu: SOEST Publication 96-06, JIMAR Contribution 96-303.

Hensel, C. 1996. *Telling Ourselves: Ethnicity and Discourse in Southwestern Alaska.* New York: Oxford University Press.

Hensel, C. 1997. Draft Research Report to the Oceans Studies Board, NRC: Bigboat Village and Training Village: A study of the effects of the Community Development Quota program in two Southwestern Alaska Villages. Unpublished.

Hensley, R., and Sherwood, T. 1993. An overview of Guam's inshore fishery. *Marine Fisheries Review* 55:129-138.

Hoag, Stephen H., Gordan J. Peltonen, and Lauri L. Sadorus. 1993. Regulations of the Pacific Halibut Fishery, 1977-1992. International Pacific Halibut Commission Technical Report No. 27.

International Pacific Halibut Commission (IPHC). 1997. 1996 Annual Report. Seattle, Washington.

Irwin, Mike (Department of Community and Regional Affairs) personal communication to Clarence Pautzke (North Pacific Fishery Management Council). 23 September, 1997.

Iversen, R., Dye, T., and Paul, L. 1989. Rights of Native Fishermen with Specific Regard to Harvesting of Bottomfish in the Northwestern Hawaiian Islands and with Regard to Harvesting of Bottomfish, Crustaceans, Precious Corals, and Open-Ocean Fish in Offshore Areas Surrounding the Entire Hawaiian Chain. Honolulu: Western Pacific Regional Fishery Management Council.

Jones, D. 1976. *Aleuts in Transition: A comparison of two villages.* Seattle, WA: University of Washington Press.

Jones, D. 1980. *A Century of Servitude: Pribilof Aleuts Under US Rule.* Washington, DC: University Press of America.

Jorgensen, J.G. 1990. *Oil Age Eskimos.* Berkeley: University of California Press.

Kingsolving, A. 1996. A Preliminary Analysis of the Quantity of "Leakage" Fish Entering American Samoa During 1995 and Their Economic Impact. Pago Pago: Unpublished report to Department of Marine and Wildlife resources.

Kinoshita, R.K. , A. Grieg , D. Colpo, and J.M. Terry. 1997. Economic Status of the Groundfish Fisheries off Alaska , 1995. Technical Report. NOAA-TM-NMFS-AFSC-72. Alaska Fisheries Science Center. Seattle, Washington.

Kohlhoff, D. 1995. *When the wind was a river: Aleut Evacuation in World War II.* Seattle, WA: University of Washington Press in association with the Aleutian/Pribilof Islands Association.

Kruse, J.A. 1986. "Subsistence and the North Slope Inupiat: The Effects of Energy Development," in Steve Langdon, ed., *Contemporary Alaskan Native Economics.* Lanham, MD: University Press of America, pp. 121-152.

Landen, M.G., Beller, M., Funk, E., Propst, M., Middaugh, J., Moolenaar, R.L. 1997. Alcohol-related injury death and alcohol availability in remote Alaska. *JAMA*, 278: 1755-1758.

Langdon, S.J. 1982. Anthropology and Alaskan fisheries management policy. *Practicing Anthropology*. 5(1):15-17.

Langdon, S.J. 1986. "Contradictions in Alaskan Native Economy and Society," in Steve Langdon, ed., *Contemporary Alaskan Native Economics*. Lanham, MD: University Press of America, pp. 29-46.

Langdon, S.J. 1987a. The Native People of Alaska. 3rd edition. Anchorage, Alaska: Greatland Graphics Press.

Langdon, S.J. 1987b. Commercial Fisheries: Implications for western Alaska Development. In T. Lane (ed.) *Developing America's Northern Frontier*. Lanham, MD: University Press of America.

Langdon, S.J. 1991. The Integration of Cash and Subsistence in Southwest Alaskan Yup'ik Eskimo Communities. In T. Matsuyama and N. Peterson (eds.) *Cash, Commoditisation and Changing Foragers*. Senri Publication No. 30. Osaka, Japan: National Museum of Ethnology.

Laughlin, W. 1980. Aleuts: Survivors of the Bering Land Bridge. New York: Holt, Rinehart and Winston.

Lonner, T.D. 1986. "Subsistence as an Economic System in Alaska: Theoretical Observations and Management Implications," in Steve Langdon, ed., *Contemporary Alaskan Native Economics*. Lanham, MD: University Press of America, pp. 15-28.

Low, L.L., G.K. Tanonaka and H.H. Shippen. 1976. Sablefish of the Northeastern Pacific Ocean and Bering Sea. Northwest Fisheries Center, National Marine Fisheries Service Processed Report October, 1976.

MSFCMA (Magnuson-Stevens Fishery Conservation and Management Act of 1996). 1996. Public Law 297, 104th Congress, 2nd session. 19 September 1996.

Marshall, D. 1988. The Alaska Economy: Performance Report, 1987. Juneau, Alaska: Department of Commerce and Economic Development. December, 1988.

Mayo, L.W. 1992. The Militarization of Guam Society in B. Robillard (ed). *Social Change in the Pacific Islands*. New York: Kegan Paul International.

McCall, A.D. 1990. Dynamic Geography of Marine Fish Populations. Seattle, Washington. University of Washington Press.

McPhetres, S.F. 1992 Elements of Social Change in the Contemporary Northern Mariana Islands in B. Robillard (ed). New York: Kegan Paul International.

Meyer Resources Inc. 1987. A report on Resident Fishing in the Hawaiian Islands. Honolulu: Southwest Fisheries Science Center, NMFS/NOAA. Wdmin. Rep. H87-8C.

Morrow, P. 1984. It is time for Drumming: A Summary of Recent Research on Yup'ik Ceremonialism. *Etudes/Inuit/Studies*. 8 (special issue):113-140.

Myers, R.F. 1993. Guam's small-boat-based fisheries. *Marine Fisheries Review*. 55:117-128.

National Research Council (NRC). 1996. Changing Numbers, Changing Needs: American Indian Demography and Public Health. Washington, D.C. National Academy Press.

Nelson, M. 1969. Bristol Bay king, chum, pink and coho salmon, 1968 : a compilation of catch and escapement data. State of Alaska, ADF&G. Informational leaflet. No. 28. March 21, 1969.

North Pacific Fishery Management Council (NPFMC). 1989a. Longline and Pot Gear Sablefish Management in the Gulf of Alaska and the Bering Sea/Aleutian Islands. Draft Supplemental Environmental Impact Statement and Regulatory Impact Review/Initial Regulatory Flexibility Analysis to the Fishery Management Plans for the Gulf of Alaska and the Bering Sea/Aleutian Islands. North Pacific Fishery Management Council, November 16, 1989.

North Pacific Fishery Management Council (NPFMC). 1989b. Fishery Management Plan for the Commercial King and Tanner Crab Fisheries in the Bering Sea/Aleutian Islands, North Pacific Fishery Management Council, January 24, 1989. Anchorage, Alaska.

North Pacific Fishery Management Council (NPFMC). 1990. North Pacific Fishery Management Council Newsletter, No. 6-90, December, 1990. Anchorage, Alaska.

North Pacific Fishery Management Council (NPFMC). 1992. Final Supplemental Environmental Impact Statement/Environmental Impact Statement for the Individual Fishing Quota Management Alternative for Fixed Gear Sablefish and Halibut Fisheries, Gulf of Alaska and Bering Sea/Aleutian Islands. North Pacific Fishery Management Council, September 15, 1992.

North Pacific Fishery Management Council (NPFMC). 1997a.. Draft Supplemental Analysis of Final License Limitation Alternative for the Groundfish Fisheries of the Bering Sea/Aleutian Islands and Gulf of Alaska and the King and Tanner Crab Fisheries of the Bering Sea/Aleutian Islands. North Pacific Fishery Management Council, May 27, 1997. Anchorage, Alaska

North Pacific Fishery Management Council (NPFMC). 1997b. Draft Environmental Assessment/ Regulatory Review for a Regulatory Amendment for Creating and Defining a Halibut Subsistence/Personal Use Fishery Category. North Pacific Fishery Management Council, May 19, 1997. Anchorage, Alaska.

North Pacific Fishery Management Council (NPFMC). 1998. Development of the Individual Fishing Quota Program for Sablefish and Halibut Longline Fisheries off Alaska. Presentation to the National Research Council's Committee to Review Individual Fishing Quota. September 4, 1998. Anchorage, Alaska. (Web http://www.noaa.gov/npfmc/ifqpaper.htm)

NRC. 1996. The Bering Sea Ecosystem. National Academy Press, Washington, DC.

Oswalt, W. 1990. Bashful No Longer: An Alaskan Eskimo Ethnohistory, 1778-1988. Norman, OK: University of Oklahoma Press.

Otto, R.S. 1981. Eastern Bering Sea Crab Fisheries, pp. 1037-1068 In: D.W. Hood and J.A. Calder (eds.) The Eastern Bering Sea Shelf: Oceanography and Resources. Vol. 2. Office of Marine Pollution Assessment. Juneau, Alaska.

Pete, M.C. 1995. Alaska's Community Development Quota Program: Community Awareness and Response, A Report of Research Findings. Western Alaska Fisheries Development Association and Bering Sea Fishermen's Association. Anchorage, Alaska.

Petterson, J.S., L.A. Palinkas, B.M. Harris, K. Barlow, and M. Downs. 1984. Sociocultural/Socioeconomic Organization of the Bristol Bay: Regional and Subregional Analyses. Social and Economic Studies Program. Technical Report No. 103. Prepared for Minerals Management Service, Anchorage. Alaska Outer Continental Shelf Office. Impact Assessment, Inc., La Jolla, California.

Pérez de Cuéllar, J. 1995. Our Creative Diversity: Report of the World Commission on Culture and Development. Paris: UNESCO Publishing.

Polovina, J.J., Mitchum, G.T., Graham, N.E., Craig, M.P., Demartini, E.E., Flint, E.N. 1994. Physical and biological consequences of a climate event in the central North Pacific. Fisheries Oceanography. 3:15-21. Honolulu: Southwest Fisheries Science Center, NMFS/NOAA.

Ray, D.S. 1975. The Eskimo of the Bering Sea: 1650-1898. Seattle: University of Washington Press.

Restrictive Access Management (RAM). 1998. The IFQ program: 1998 report to the fleet. NMFS/ FAKR/RAM. Juneau, Alaska.

Rigby, P. 1984. Alaska domestic groundfish fishery for the years 1970 through 1980 with a review of two historic fisheries—Pacific cod (Gadus macrocephalus) and sablefish (Anoplopoma fimbria). State of Alaska, ADF&G, Division of Commercial Fisheries. Technical Report No. 108. Juneau, Alaska.

Sahlins, M. 1993. Good-bye to Tristes Tropes: Ethnography in the Context of Modern World History. Journ. of Mod. Hist. 65:1-25.

Sambrotto, R.N., Goering, J.J., McRoy, C.P. 1986. Large yearly production of phytoplankton in the western Bering Strait. Science. 225:1147-1151.

Seafood Business. 1988. Americans take over the zone. Seafood Business. 7:98-102.

Severance, C., and Franco, R. 1989. Justification and Design of Limited Entry Alternatives for the Offshore Fisheries of American Samoa, and an Examination of Preferential Fishing Rights for the Native People of American Samoa within a Limited Entry Context. Honolulu: Western Pacific Fishery Management Council.

Shaw, R. 1983. Archaeology of the Manokinak Site: A study of the cultural transition between late Norton and historic Eskimo. Unpublished doctoral thesis. Washington State University. Pullman.

State of Alaska. 1995. Revised Draft Report: Economic Impacts of the Pollock Community Development Quota Program. State of Alaska. Anchorage, Alaska.

Thomas, N. 1991. Entangled Objects. Harvard University Press.

Thompson, W. and Freeman, N. 1930. History of the Pacific Halibut Fishery. Wrigley Printing Co., Ltd. Vancouver, B.C., Canada. Thompson and Freeman.

Townsend, R., and Pooley, S. Fractional licenses: an alternative to license buy-backs. *Land Economics*. 71:10-25.

United States District Court, District of Alaska. March, 1981. Bristol Bay Herring Marketing Cooperative, an Alaska Corporation, and Basil Backford, and Nels Franklin, and Western Alaska Cooperative Marketing Association, an Alaska Corporation, and Harold Samuelson, Jr., and Joe McGill vs. Ronald Skoog, Commissioner, Alaska Department of Fish and Game, and Alaska Board of Fisheries, and the State of Alaska. Case No. A81-043 Civ. Unpublished decision.

Van Stone, J.W. 1960. "A Successful Combination of Subsistence and Wage Economics on the Village Level," *Economic Development and Cultural Change*. 8:174-191.

Van Stone, J.W. 1962. *Point Hope: An Eskimo Village in Transition*. Seattle: University of Washington Press.

Van Stone, J.W. 1967. Eskimos of the Nushagek River: An Ethnographic History. Seattle: University of Washington Publications in Anthropology 15.

Van Stone, J.W. 1984. "Mainland Southwest Alaska Eskimo" in Handbook of Native American Indians: Volume 5 Arctic. David Damas ed. Washington: Smithsonian Institution Press. pp. 224-242.

Walker, J. 1996. Sociology of Hawaii Charter Boat Fishing. Honolulu: UH/NOAA/JIMAR Contribution 96-309.

Western Pacific Regional Fishery Management Council (WPRFMC). 1997a. The Value of the Fisheries in the Western Pacific Regional Fishery Management Council's Area. Honolulu.

Western Pacific Regional Fishery Management Council (WPRFMC) 1997b. Pelagic Fisheries of the Western Pacific Region, 1996 Annual Report. Honolulu.

Wolfe, R.A. 1980. Food production in a western Eskimo population. Ann Arbor: University of Michigan Press.

Wolfe, R.A. 1984. Commercial Fishing in the hunting-gathering economy of a Yukon River Yup'ik society. *Etudes/Inuit/Studies*. 8 (special issue):159-183.

Wolfe, R.A. 1986. "The Economic Efficiency of Food Production in a western Alaska Eskimo Population," in Steve Langdon, ed., *Contemporary Alaskan Native Economics*. Lanham, MD: University Press of America, pp. 101-120.

Zahn, M. 1970. Japanese tanner crab fishery in the Eastern Bering Sea. Commercial Fisheries Review. 32:52-56.

Appendixes

Appendix A

Authorizing Legislation

**MAGNUSON-STEVENS FISHERY CONSERVATION AND
MANAGEMENT ACT
PUBLIC LAW 104-297 (SEC. 108(H)).**

(h) COMMUNITY DEVELOPMENT QUOTA REPORT— Not later than October 1, 1998, the National Academy of Sciences, in consultation with the Secretary, the North Pacific and Western Pacific Councils, communities and organizations participating in the program, participants in affected fisheries, and the affected States, shall submit to the Secretary of Commerce and Congress a comprehensive report on the performance and effectiveness of the community development quota programs under the authority of the North Pacific and Western Pacific Councils. The report shall—

(1) evaluate the extent to which such programs have met the objective of providing communities with the means to develop ongoing commercial fishing activities;

(2) evaluate the manner and extent to which such programs have resulted in the communities and residents—

(A) receiving employment opportunities in commercial fishing and processing; and

(B) obtaining the capital necessary to invest in commercial fishing, fish processing, and commercial fishing support projects (including infrastructure to support commercial fishing);

(3) evaluate the social and economic conditions in the participating communities and the extent to which alternative private sector employment opportunities exist;

(4) evaluate the economic impacts on participants in the affected fisheries, taking into account the condition of the fishery resource, the market, and other relevant factors;

(5) recommend a proposed schedule for accomplishing the developmental purposes of community development quotas; and

(6) address such other matters as the National Academy of Sciences deems appropriate.

Appendix B

Biographical Sketches of the Committee's Members

John E. Hobbie earned his Ph.D. in zoology from Indiana University. Dr. Hobbie is a senior scientist at the Marine biological Laboratory in Woods Hole and is co-director of its Ecosystem Center. Dr. Hobbie leads the Land-Margin Ecosystem Research program of the National Science Foundation. His research interests include arctic and antarctic limnology, estuarine ecology, and the global carbon cycle. Dr. Hobbie has been a member of the Ocean Studies Board since 1995.

Daniel W. Bromley earned his Ph.D. in natural resource economics from Oregon State University. Dr. Bromley is Anderson-Bascom Professor and Chair, Department of Agricultural and Applied Economics at the University of Wisconsin, Madison. His research interests include institutional economics, political economy, natural resource economics, and the environmental implications of economic development. He has a specific interest in common property resource management—he was the chair of the NRC Panel on Common Property Resource Management, and he served as president of the International Association for the Study of Common Property.

Paul K. Dayton earned his Ph.D. from the University of Washington, Seattle. Dr. Dayton has been a professor at the Scripps Institution of Oceanography, University of California, San Diego, since 1982. His major research interest is in the field of marine ecology.

Daniel D. Huppert earned his Ph.D. from the University of Washington. Dr. Huppert was an economist and program leader for the National Marine Fisheries

Service in San Diego, and then became an associate professor at the University of Washington's School of Marine Affairs in 1989. Dr. Huppert served on Scientific and Statistical Committees for the Pacific and North Pacific Fishery Management Councils for 15 years, and he is currently on the Independent Economic Analysis Board for the Northwest Power Planning Council. His research interests include endangered species planning, commercial fishery regulation, coastal ecosystems conservation and planning, and international management of Pacific salmon.

Stephen J. Langdon earned his Ph.D. in anthropology from Stanford University. Dr. Langdon has been a professor in the Department of Anthropology at the University of Alaska, Anchorage, since 1987. His research interests include ecological and economic anthropology. His research focuses on the maritime societies of the Northwest coast of the United States.

Seth Macinko earned his Ph.D. from the University of California, Berkeley. Dr. Macinko has been a social and economic policy analyst at the Alaska Department of Fish and Game since 1993. His research focuses on institutional analysis of natural resource use and management. He fished commercially off Alaska from 1979 to 1983.

Marshall D. Sahlins earned his Ph.D. in anthropology from Columbia University. Dr. Sahlins is the Charles F. Grey Distinguished Service Professor at the University of Chicago. His research interests include the cultural processes of historical change. Dr. Sahlins is a member of the National Academy of Sciences.

Craig J. Severance earned his Ph.D. in anthropology from the University of Oregon. Dr. Severance is a professor at the University of Hawaii at Hilo, with expertise in maritime cultures of the Western Pacific. He is a member of the Scientific and Statistical Committee (SSC) of the Western Pacific Regional Fishery Management Council and a part-time recreational/commercial fisherman.

Ronald L. Trosper earned his Ph.D. in economics from Harvard University. Dr. Trosper has been the Director of the Native American Forestry Program at Northern Arizona University since 1989. His research focuses on issues of economic development and ecosystem management by Native American peoples. He is a member of the Confederated Salish and Kootenai tribes of the Flathead Indian Reservation in Montana.

Miranda Wright earned her Master's degree in anthropology from the University of Alaska, Fairbanks. Ms. Wright serves as Executive Director of the Doyon Foundation, a non-profit organization which focuses on education and professional opportunities for Alaska Natives. She is an anthropologist with research interests in the areas of social and economic pressures on the indigenous peoples of the Alaskan subarctic.

Appendix C

Acknowledgments

The preparation of a report such as this one takes input from many people. The Committee wishes to extend its sincere appreciation to all the people who shared their expertise and experience with us during the study process. This includes the North Pacific and Western Regional Pacific Fishery Management Councils, the National Marine Fisheries Service, the State of Alaska, the State of Hawaii, and the Community Development Quota Associations. We would also like to thank the many people who participated in our information-gathering sessions, led us on field trips, helped with our research, and contributed to the study in other ways. In particular, we would like to thank the following people:

NORTH PACIFIC FISHERY MANAGEMENT COUNCIL REGION

Bob Alverson, Fishing Vessel Owners Association, Seattle
Julie Anderson, Department of Community and Regional Affairs, Juneau
Eugene Asicksik, Norton Sound Economic Development Corporation, Anchorage
David Benton, Department of Fish and Game, Juneau
Sally Bibb, National Marine Fisheries Service, Juneau
David Bill, Coastal Villages Fisheries Cooperative, Toksook Bay
John Bundy, Glacier Fish Company, Seattle
Al Burch, Alaska Draggers Association, Kodiak
Jeff Bush, Department of Commerce and Economic Development, Juneau
Tom Casey, Alaska Fisheries Conservation Group, Woodinville
Bill Charles, Yukon Delta Fisheries Development Association, Emmonak
Norman Cohen, Coastal Villages Fisheries Cooperative, Juneau

Larry Cotter, Aleutian Pribilof Island Community Development Association, Juneau
Lemar Cotton, Department of Community and Regional Affairs, Juneau
Mark Denison, Icicle Seafoods Inc., Seattle
Lou Flemming, Golden Alaska Seafoods, Seattle
William Fox, National Marine Fisheries Service, Silver Spring, Maryland
Rodney Fujita, Environmental Defense Fund, Oakland, California
Steve Ganey, Alaska Marine Conservation Council, Anchorage
Amy Gautam, National Marine Fisheries Service, Silver Spring, Maryland
John Gauvin, Groundfish Forum, Seattle
Jay Ginter, National Marine Fisheries Service, Juneau
Ed Glotfelty, Yukon Delta Fisheries Development Association, Seward
Glenn Haight, Department of Community and Regional Affairs, Juneau
Marcus Hartley, Bering Sea Fishermen's Association, Anchorage
Chase Hensel, Consultant to the Committee
Ralph Hoard, Icicle Seafoods Inc., Seattle
Jan Jacobs, American Seafoods Company, Seattle
Patrick Kelly, Norton Sound Seafood Products, Seattle
Joe Kyle, Aleutian Pribilof Island Community Development Association, Juneau
Richard Lauber, North Pacific Fishery Management Council, Anchorage
Dan Leston, Sealaska Corporation, Seattle
Robert Loescher, Sealaska Corporation, Juneau
Mary McBurney, Western Alaska Fisheries Development Association, Anchorage
Carl Merculief, Central Bering Sea Fishermen's Association, Saint Paul Island
Larry Merculieff, Central Bering Sea Fishermen's Association, Anchorage
John Moller, Aleutian Pribilof Island Community Development Association, Juneau
Judith Nelson, Bristol Bay Economic Development Corporation, Dillingham
Kris Norosz, Petersburg Fisheries, Petersburg
Karl Ohls, Department of Commerce and Economic Development, Juneau
Bill Orr, Golden Age Fisheries, Seattle
Brent Paine, United Catcher Boats, Seattle
Clarence Pautzke, North Pacific Fishery Management Council, Anchorage
Wally Pereyra, Arctic Storm Inc., Seattle
Paul Peyton, Fisheries Business Consultant, Juneau
Dennis Phelan, Pacific Seafood Processors Association, Washington, D.C.
Evelyn Pinkerton, Simon Fraser University, British Columbia
Joe Plescha, Trident Seafoods, Seattle
Thomas Punguk, Fisherman, Golovin
Mike Rink, Fisherman, Nome
Susan Sabella, Greenpeace, Washington, D.C.
Gilda Shellikof, Aleutian Pribilof Island Community Development Association, Juneau
Tim Smith, Nome Fishermen's Association, Nome
Rick Steiner, University of Alaska Marine Advisory Program, Anchorage
Tom Suryan, Skippers for Equitable Access, Seattle

Arni Thompson, Alaska Crab Coalition, Seattle
Tim Towarak, Office of the Governor, Juneau
Richard Tremaine, E3 Consulting Services, Anchorage
Jon Zuck, Consultant, Anchorage

WESTERN PACIFIC REGION

William Aila, Fisherman, Waianae
Paul Bartram, Akala Products Inc., Honolulu
Ray Clarke, National Marine Fisheries Service, Southwest Region, Honolulu
Carl Christensen, Native Hawaiian Legal Corporation, Honolulu
Manuel Duenas, Guam Fisherman's Cooperative, Guam
Shane Jones, Commissioner, Treaty of Waitangi Fisheries Commission, Auckland
Isaac D. Harp, Fisherman, Lahaina
Mark Hodges, Kahoolawe Island Reserve Commission, Honolulu
David Kaltoff, Saint Peter Fishing, Honolulu
Buddy Keala, Hui O Kuapa, Kaunakakai
Patrick Kirch, University of California, Berkeley
Wayde Lee, Moomomi Demonstration Project, Kaunakakai
Kimberly Lowe, Department of Land and Natural Resources, Honolulu
Collette Machado, Office of Hawaiian Affairs, Honolulu
Linda Paul, Hawaii Audubon Society, Honolulu
Walter Ritte, Native Practitioner, Kaunukakai
Don Schug, Western Pacific Fishery Management Council, Honolulu
Richard Seman, Commonwealth Northern Mariana Islands, Division of Fish and
    Wildlife, Saipan
Kitty Simonds, Western Pacific Fishery Management Council, Honolulu
Brooks Takenaka, United Fish Agency Ltd., Honolulu
Representative David A. Tarnas, Honolulu
Mililani Trask, Ka Láhui Hawaii, Hilo
David Ziemann, Oceanic Institute, Waimanalu

Appendix D

State of Alaska CDQ Regulations

PERMANENT REGULATIONS
TITLE 05
CHAPTER 039
ARTICLE 06

5 AAC 39.690

BERING SEA/ALEUTIAN ISLANDS KING AND TANNER CRAB
COMMUNITY DEVELOPMENT QUOTA (CDQ) FISHERIES
MANAGEMENT PLAN.

(a) Male red and blue king crab and male Tanner crab may be taken in a Bering Sea/Aleutian Islands CDQ fishery only under the conditions of a permit issued by the commissioner. Female crab of these species may not be taken.

(b) In the Bering Sea/Aleutian Islands CDQ fishery

(1) male king crab may be taken only in Registration Area T as described in 5 AAC 34.800 and Registration Area Q as described in 5 AAC 34.900 ; and

(2) male Tanner crab may be taken only in the Bering Sea District of Registration Area J as described in 5 AAC 35.505 (e).

(c) Male king crab and male Tanner crab may be taken only with pots.

(d) Unless otherwise specified in a permit issued under (e) of this section, an operator of a vessel fishing CDQ crab allocations shall comply with all regulations in 5 AAC 34, 5 AAC 35, and 5 AAC 39 that are applicable to the area and species of crab being fished.

(e) In the permit required in (a) of this section the commissioner may, as the commissioner determines it necessary for the conservation and management of the resource,

(1) specify the species of king or Tanner crab that may be taken;

(2) specify an area or areas open to CDQ fishing operations;

(3) specify registration requirements;

(4) specify fishing periods;

(5) specify gear requirements, including the numbers of pots;

(6) designate the poundage of the federal CDQ allocation of a species of king or Tanner crab that may be taken by a CDQ group in a registration area or portion of a registration area;

(7) if a CDQ fishery is conducted immediately before the opening of the commercial fishery for a species of king or Tanner crab, limit a CDQ fishery harvest, by establishing fishing periods and the amount of the CDQ allocation for that same species of king or Tanner crab that may be taken before the opening of the commercial fishery;

(8) require onboard observers during fishing operations;

(9) require the operator of a vessel fishing CDQ crab allocations to notify the department of fishing time, delivery time, and delivery destination;

(10) establish reporting requirements;

(11) require logbooks;

(12) establish times and areas allowed for placement and removal of gear;

(13) establish times, areas, and conditions allowed for storage of gear;

(14) set out other conditions deemed necessary by the commissioner.

(f) A permit issued under (e) of this section will be considered the registration required by 5 AAC 34.020 or 5 AAC 35.020 .

(g) Notwithstanding 5 AAC 34.005 and 5 AAC 35.005 , participation by a CDQ permit holder in an exclusive or superexclusive king or Tanner crab fishery does not preclude a vessel or permit holder from participation in a CDQ crab fishery.

(h) Notwithstanding 5 AAC 34.005 and 5 AAC 35.005 , participation by a CDQ permit holder in a CDQ king or Tanner crab fishery does not preclude a vessel or permit holder from participation in an exclusive or superexclusive crab fishery.

History -

Eff. 9/12/97, Register 143

Authority -

AS 16.05.251

6 AAC 93.015

DELEGATION OF AUTHORITY.

The commissioners of the Department of Community and Regional Affairs, Department of Fish and Game, and Department of Commerce and Economic Development, acting jointly, are the governor's designees for the purposes of this chapter. The governor's designees shall (1) solicit community development plan (CDP) applications from eligible communities; (2) conduct the initial review and evaluation of proposed CDPs; (3) make recommendations for community development quota (CDQ) allocations to the governor; and (4) review and recommend for approval amendments to existing CDPs. The governor will make all final recommendations regarding CDP applications and CDQ allocations to the United States Secretary of Commerce (secretary) in accordance with 50 C.F.R. 675.27(a) - (c), 50 C.F.R. 676.24(c), and this chapter.

History -

Eff. 11/18/92, Register 124; am 4/10/93, Register 126; am 8/13/94, Register 131

Authority -

Art. III, sec. 1, Ak. Const.

Art. III, sec. 24, Ak. Const.

6 AAC 93.025

COMMUNITY DEVELOPMENT QUOTA ALLOCATIONS.

(a) To receive a CDQ allocation, a qualified applicant must submit to the Office of the Governor a complete CDP application before the end of the CDP application period established in 6 AAC 93.020 (a). Except for exceptional circumstances beyond the control of the qualified applicant, the governor's designees will not review a late CDP application. A CDP application is complete if the CDP application includes:

(1) for a pollock CDQ allocation,

(A) the information described at 50 C.F.R. 675.27(b), as amended as of July 14, 1994;

(B) a statement from the applicant that the applicant is a "qualified applicant" as defined at 50 C.F.R. 675.27(d)(6), as amended as of July 14, 1994; and

(C) a statement from the applicant that each community participating in the CDP application is an eligible community as described at 50 C.F.R. 675.27(d)(2), as amended as of July 14, 1994;

(2) for a halibut CDQ allocation or a sablefish CDQ allocation,

(A) the information described at 50 C.F.R. 676.24(d), as amended as of July 14, 1994;

(B) a statement from the applicant that the applicant is a "qualified applicant" as defined at 50 C.F.R. 676.24(f)(6), as amended as of July 14, 1994; and

(C) a statement from the qualified applicant that each community participating in the CDP application is an eligible community as described at 50 C.F.R. 676.24(f)(2), as amended as of July 14, 1994;

(3) a list of the communities participating in the CDP application; and

(4) a statement of support from the governing body of each community participating in the CDP application.

(b) An eligible community may not concurrently apply for or receive more than one CDQ allocation during a single CDP application period for each of the following species:

(1) pollock;

(2) halibut;

(3) sablefish.

(c) If the applicant for a CDQ is a managing organization, that organization must have

(1) a board of directors whose membership is composed of at least 75 percent resident fishermen from the community or group of communities involved in the CDP application, with at least one member from each participating community; and

(2) a statement of support from each community on whose behalf the organization is applying, that was approved by the governing body of that community.

(d) If a managing organization will participate in the fishery on behalf of an applicant, but is not the applicant, that organization must

(1) provide a statement of support from the governing body of each community that the organization represents; and

(2) document the legal relationship between the applicant and the managing organization, through a contract or other legally binding agreement, that clearly describes the responsibilities and obligations of the parties.

(e) In addition to the information required under (a) - (c) of this section, 50 C.F.R. 675.27, and 50 C.F.R. 676.24 a qualified applicant shall provide with the CDP application all information regarding the particular benefits that a CDQ allocation under the CDP would generate for the Bering Sea/Aleutians Island region, the state, or the United States.

(f) The requirement for a statement of support under (a), (c)(2), or (d)(1) of this section may be satisfied by providing a copy of a resolution, letter, or other appropriate expression of support from the governing body of that community.

(g) In addition to the information required in this section, a qualified applicant shall submit to the Office of the Governor additional information that the governor or the governor's designees determine to be necessary to determine whether to recommend the complete CDP application to the secretary for approval.

History -

Eff. 11/18/92, Register 124; am 4/10/93, Register 126; am 8/13/94, Register 131

Authority -

Art. III, sec. 1, Ak. Const.

Art. III, sec. 24, Ak. Const.

6 AAC 93.040

FINAL EVALUATION OF COMPLETE CDP APPLICATIONS.

(a) Following the close of the CDP application period, the governor's designees shall evaluate all complete CDP applications for a

(1) pollock CDQ allocation to determine if they meet the requirements of 50 C.F.R. 675.27, as amended as of July 14, 1994, and this chapter; and

(2) halibut or sablefish CDQ allocation to determine if they meet the requirements of 50 C.F.R. 675.24, as amended as of July 14, 1994, and this chapter.

(b) The governor's designees shall consider the following factors when reviewing a complete CDP application:

(1) the number of eligible communities participating in the CDQ program;

(2) the size of the allocation of fishery resource requested by the qualified applicant and the number of years the qualified applicant requires the allocation to achieve the milestones, goals and objectives of the CDP as stated in the complete CDP application;

(3) the degree to which the project is expected, if any, to develop a self-sustaining local fisheries economy, and the proposed schedule for transition from reliance on a CDQ allocation to economic self-sufficiency;

(4) the degree to which the project is expected, if any, to generate capital or equity in the local fisheries economy or infrastructure, or investment in commercial fishing or fish processing operations;

(5) the contractual relationship between the qualified applicant and joint venture partners, if any, and the managing organization.

(c) For each complete CDP application evaluated by the governor's designees, the governor will make a written finding that the complete CDP application

(1) satisfies the requirements of 50 C.F.R. 675.27 or 50 C.F.R. 676.24, as amended as of July 14, 1994, and this chapter, and will be recommended to the secretary for approval for a CDQ allocation in the amount requested by the qualified applicant;

(2) satisfies the requirements of 50 C.F.R. 675.27 or 50 C.F.R. 676.24, as amended as of July 14, 1994, and this chapter, and will be recommended to the secretary for approval with a reduced CDQ allocation from the amount requested by the qualified applicant; or

(3) does not satisfy the requirements of 50 C.F.R. 675.27 or 50 C.F.R. 676.24, as amended as of July 14, 1994, and this chapter, and will not be recommended to the secretary for approval.

(d) If there is a sufficient quota of fishery resource available to meet the CDQ allocations requested in all of the complete CDP applications, the governor will, in the governor's discretion, recommend all of the complete CDP applications that satisfy the requirements of 50 C.F.R. 675.27 or 50 C.F.R. 676.24, as amended as of July 14, 1994, and this chapter to the secretary for approval.

(e) If there is an insufficient quota of fishery resource available to meet the combined total CDQ allocations requested in all of the complete CDP applications that satisfy the requirements of 50 C.F.R. 675.27 or 50 C.F.R. 676.24, as amended as of July 14, 1994, and this chapter, the governor will, in the governor's discretion and after consultation in accordance with (f) of this section,

(1) apportion the available quota among the qualified applicants and recommend the apportionment to the secretary for approval; or

(2) select those complete applications that the governor believes best satisfy the objectives, requirements, and criteria of the CDQ program and recommend those applications to the secretary for approval; a recommendation under this paragraph may also include a recommendation for an apportionment in accordance with (1) of this subsection.

(f) Before the governor recommends an apportionment of the quota under (e) of this section, the governor will consult with the qualified applicants that may be affected by the proposed apportionment. The governor will, in the governor's discretion, request a qualified applicant to submit a revised complete CDP application to assist the governor in determining the

(1) economic feasibility and likelihood of success of the CDP with an allocation of fishery resource less than that requested in the complete CDP application; and

(2) particular benefits that may be derived by participating eligible communities affected by an allocation of fishery resource less than that requested in the complete CDP application.

(g) Priority in apportioning the quota of fishery resource under (e) of this section will be based upon maximizing the benefits of the CDP program to the greatest number of participating eligible communities.

(h) Before forwarding recommendations to the secretary under 6 AAC 93.045 , the governor, or the governor's designees, will consult with the North Pacific Fishery Management Council (council) regarding the complete CDP applications to be recommended by the governor for CDQ allocations and will incorporate any comments from the council into the written findings required under (c) of this section and 50 C.F.R. 675.27(b) or 50 C.F.R. 676.24(d), as amended as of July 14, 1994.

History -

Eff. 11/18/92, Register 124; am 4/10/93, Register 126; am 8/13/94, Register 131

Authority -

Art. III, sec. 1, Ak. Const.

Art. III, sec. 24, Ak. Const.

6 AAC 93.045

RECOMMENDATIONS TO THE SECRETARY.

Within five working days after the end of the review and evaluation period established under 6 AAC 93.020 , the governor will

(1) forward to the secretary written recommendations for approval of CDP application and CDQ allocations; and

(2) notify in writing every CDP applicant whether the applicant's CDP was recommended to the secretary, including any reduction of allocation made under 6 AAC 93.040 .

History -

Eff. 11/18/92, Register 124; am 4/10/93, Register 126

Authority -

Art. III, Sec. 1, Ak. Const.

Art. III, Sec. 24, Ak. Const.

6 AAC 93.055

AMENDMENTS TO A CDP AND REQUESTS FOR INCREASE IN ALLOCATION.

(a) A qualified applicant that seeks to amend a complete CDP application under 50 C.F.R. 675.27(e)(3) or 50 C.F.R. 676.24(g)(3), as amended as of July 14, 1994, shall submit to the Office of the Governor a written request for approval of the amendment. The governor or the governor's designees will recommend to the secretary for approval a request to amend a complete CDP application under

(1) 50 C.F.R. 675.27(e)(3), as amended as of July 14, 1994, if the CDP, if changed, would continue to meet the requirements under 50 C.F.R. 675.27, as amended as of July 14, 1994; and

(2) 50 C.F.R. 676.24(g)(3), as amended as of July 14, 1994, if the CDP, if changed, would continue to meet the criteria under 50 C.F.R. 676.24(f), as amended as of July 14, 1994.

(b) If a qualified applicant seeks to increase its pollock CDQ allocation under a multi-year CDP, its halibut CDQ allocation, or its sablefish CDQ allocation, the qualified applicant shall submit a new complete CDP application to the Office of the Governor for approval as required under this chapter and under 50 C.F.R. 675.27 or 50 C.F.R. 676.24, as amended as of July 14, 1994.

History -

Eff. 11/18/92, Register 124; am 4/10/93, Register 126; am 8/13/94, Register 131

Authority -

Art. III, sec. 1, Ak. Const.

Art. III, sec. 24, Ak. Const.

6 AAC 93.900

DEFINITIONS.

In this chapter

(1) "allocation" means a portion of a CDQ that is allocated by the secretary to an individual qualified applicant;

(2) "CDP" or "community development plan" means a plan submitted by a qualified applicant for a CDQ allocation that meets the requirements of 50 C.F.R. 675.27 or 50 C.F.R. 676.24, as amended as of July 14, 1994;

(3) "CDQ" or "community development quota" means a specific quota of fishery resource set aside for community development purposes by the council as part of the Western Alaska Community Development Quota Program established under 50 C.F.R. 675.27 or 50 C.F.R. 676.24, as amended as of July 14, 1994;

(4) "council" means the North Pacific Fishery Management Council established in 16 U.S.C. 1852, as amended as of July 14, 1994;

(5) "eligible community" means a community that

(A) for a pollock CDQ, meets the criteria of or is listed in Table 1 of 50 C.F.R. 675.27(d)(2), as amended as of July 14, 1994; or

(B) for a halibut CDQ or a sablefish CDQ, meets the criteria of or is listed in or is listed in Table 1 of 50 C.F.R. 676.24, as amended as of July 14, 1994;

(6) "governing body of an eligible community" means a city council, traditional council, or Indian Reorganization Act (IRA) Council of an eligible community;

(7) "managing organization" means a legally recognized corporation, company, association, or other entity responsible for the management or operation of a CDP that is able to sue, be sued, enter into binding agreements, obtain loans, and own property; a "managing organization" may be either a qualified applicant, or a

separate party operating a CDP under contract with or in partnership with a qualified applicant;

(8) "qualified applicant" means an organization described in 50 C.F.R. 675.27(d)(6) or 50 C.F.R. 676.24(f)(6), as amended as of July 14, 1994;

(9) "resident fisherman" means a resident fisherman as defined in 50 C.F.R. 675.27(d)(7) or 50 C.F.R. 676.24(f)(7), as amended as of July 14, 1994;

(10) "secretary" means the United States Secretary of Commerce.

History -

Eff. 11/18/92, Register 124; am 4/10/93, Register 126; am 8/13/94, Register 131

Authority -

Art. III, sec. 1, Ak. Const.

Art. III, sec. 24, Ak. Const.

15 AAC 77.055

REQUIREMENTS FOR NONPROFIT CORPORATIONS.

(a) As a condition to a person obtaining a credit under AS 43.77.040 , a nonprofit corporation receiving or anticipating receipt of a contribution shall separately account for the amount of the contribution and agree to allow the department to audit all accounts relating to the contributions.

(b) In order to separately account for the expenditure of the contribution, the nonprofit corporation must establish, the following special accounts, as applicable, in its accounting system:

(1) CDQ fisheries scholarships;

(2) CDQ fisheries industry training;

(3) CDQ fisheries transportation grants;

(4) CDQ fisheries transportation loans;

(5) CDQ fisheries facilities grants;

(6) CDQ fisheries facilities loans; and

(7) CDQ fisheries research grants.

(c) The use of monies in the special accounts must be determined on a first-in-first-out (FIFO) basis. All contributions, with the exception of those initially placed in a loan account, must be distributed from the credited account and used for a qualified purpose by the end of the calendar year following the calendar year of receipt. Contributions placed in a loan account may be accumulated without the requirement for distribution within a specified time period. A contribution initially accounted for in a loan account that is transferred to another account must be distributed from that account by the end of the calendar year following the calendar year of the transfer. Contributions may not be transferred from a non-loan account to a loan account.

(d) No credit is allowed, and prior credits will, in the department's discretion, be revoked, if the person, or a party related to the person receives a loan or grant from a nonprofit corporation to which that person has made or will make a contribution that qualifies for the credit, unless the person first obtains written authorization from the department to receive the loan or grant.

(e) A nonprofit corporation, to accommodate potential contributors, may apply to the department and request a determination that the intended use of contributions for a specifically designated program satisfies one or more of the requirements of AS 43.77.040 and this chapter

History -

Eff. 4/20/94, Register 130

Authority -

AS 43.77.040

AS 43.77.070

AS 43.77.200

Appendix E

Federal CDQ Regulations

CODE OF FEDERAL REGULATIONS
SEC. 679.30 GENERAL CDQ REGULATIONS.

(a) State of Alaska CDQ responsibilities

(1) Compliance. The State of Alaska must be able to ensure implementation of the CDPs once approved by NMFS. To accomplish this, the State must establish a monitoring system that defines what constitutes compliance and non-compliance.

(2) Public hearings. Prior to granting approval of a CDP recommended by the Governor, NMFS shall find that the Governor developed and approved the CDP after conducting at least one public hearing, at an appropriate time and location in the geographical area concerned, so as to allow all interested persons an opportunity to be heard. Hearing(s) on the CDP do not have to be held on the actual documents submitted to the Governor under paragraph (b) of this section, but must cover the substance and content of the proposed CDP in such a manner that the general public, and particularly the affected parties, have a reasonable opportunity to understand the impact of the CDP. The Governor must provide reasonable public notification of hearing date(s) and location(s). The Governor must make available for public review, at the time of public notification of the hearing, all state materials pertinent to the hearing(s) and must include a transcript or summary of the public hearing(s) with the Governor's recommendations to NMFS in accordance with this subpart. At the same time this transcript is submitted to

NMFS, it must be made available, upon request, to the public. The public hearing held by the Governor will serve as the public hearing for purposes of NMFS review under paragraph (c) of this section.

(3) Council consultation. Before sending his/her recommendations for approval of CDPs to NMFS, the Governor must consult with the Council, and make available, upon request, CDPs that are not part of the Governor's recommendations.

(b) CDP application. The Governor, after consultation with the Council, shall include in his or her written findings to NMFS recommending approval of a CDP, that the CDP meets the requirements of these regulations, the Magnuson Act, the Alaska Coastal Management Program, and other applicable law. At a minimum, the submission must discuss the determination of a community as eligible; information regarding community development, including goals and objectives; business information; and a statement of the managing organization's qualifications. For purposes of this section, an eligible community includes any community or group of communities that meets the criteria set out in paragraph (d) of this section. Applications for a CDP must include the following information:

(1) Community development information. Community development information includes:

(i) Project description. A description of the CDP projects that are proposed to be funded by the CDQ and how the CDP projects satisfy the goals and purpose of the CDQ program.

(ii) Allocation request. The allocation of each CDQ species requested for each subarea or district of the BSAI, as defined at Sec. 679.2 and for each IPHC regulatory area, as prescribed in the annual management measures published in the Federal Register pursuant to Sec. 300.62 of chapter III of this title.

(iii) Project schedule. The length of time the CDQ will be necessary to achieve the goals and objectives of the CDP, including a project schedule with measurable milestones for determining progress.

(iv) Employment. The number of individuals to be employed under the CDP, the nature of the work provided, the number of employee-hours anticipated per year, and the availability of labor from the applicant's community(ies).

(v) Vocational and educational programs. Description of the vocational and educational training programs that a CDQ allocation under the CDP would generate.

(vi) Existing infrastructure. Description of existing fishery-related infrastructure and how the CDP would use or enhance existing harvesting or processing capabilities, support facilities, and human resources.

(vii) New capital. Description of how the CDP would generate new capital or equity for the applicant's fishing and/or processing operations.

(viii) Transition plan. A plan and schedule for transition from reliance on the CDQ allocation under the CDP to self-sufficiency in fisheries.

(ix) Short- and long-term benefits. A description of short- and long-term benefits to the applicant from the CDQ allocation.

(2) Business information. Business information includes:

(i) Method of harvest. Description of the intended method of harvesting the CDQ allocation, including the types of products to be produced; amounts to be harvested; when, where, and how harvesting is to be conducted; and names and permit numbers of the vessels that will be used to harvest a CDQ allocation.

(ii) Target market and competition. Description of the target market for sale of products and competition existing or known to be developing in the target market.

(iii) Business relationships. Description of business relationships between all business partners or with other business interests, if any, including arrangements for management, audit control, and a plan to prevent quota overages. For purposes of this section, business partners means all individuals who have a financial interest in the CDQ project.

(iv) Profit sharing. Description of profit sharing arrangements.

(v) Funding. Description of all funding and financing plans.

(vi) Partnerships. Description of joint venture arrangements, loans, or other partnership arrangements, including the distribution of proceeds among the parties.

(vii) General budget for implementing the CDP. A general budget is a general account of estimated income and expenditures for each CDP project that is described in paragraph (b)(1)(i) of this section for the total number of calendar years that the CDP is in effect.

(viii) Capital equipment. A list of all capital equipment.

(ix) Cash flow. A cash flow and break-even analysis.

(x) Income statement. A balance sheet and income statement, including profit, loss, and return on investment for the proposed CDP.

(3) Statement of managing organization's qualifications. Statement of the managing organization's qualifications includes:

(i) Structure and personnel. Information regarding its management structure and key personnel, such as resumes and references; including the name, address, fax number, and telephone number of the managing organization's representative; and

(ii) Management qualifications. A description of how the managing organization is qualified to manage a CDQ allocation and prevent quota overages. For purposes of this section, a qualified managing organization means any organization or firm that would assume responsibility for managing all or part of the CDP and that meets the following criteria:

(A) Official letter of support. Documentation of support from each community represented by the applicant for a CDP through an official letter of support approved by the governing body of the community.

(B) Legal relationship. Documentation of a legal relationship between the CDP applicant and the managing organization (if the managing organization is different from the CDP applicant), which clearly describes the responsibilities and obligations of each party as demonstrated through a contract or other legally binding agreement.

(C) Expertise. Demonstration of management and technical expertise necessary to carry out the CDP as proposed by the CDP application (e.g., proven business experience as shown by a balance and income statement, including profit, loss, and the return on investment on all business ventures within the previous 12 months by the managing organization).

(c) Review and approval of CDPs—(1) Consistent with criteria. (i) Upon receipt by NMFS of the Governor's recommendation for approval of proposed CDPs, NMFS will review the record to determine whether the community eligibility criteria and the evaluation criteria set forth in paragraph (d) of this section have been met. NMFS shall then approve or disapprove the Governor's recommendation within 45 days of its receipt.

(ii) In the event of approval, NMFS shall notify the Governor and the Council in writing that the Governor's recommendations for CDPs are consistent with the evaluation criteria under paragraph (d) of this section and other applicable law, including NMFS reasons for approval.

(iii) Publication of the decision, including the percentage of each CDQ reserve for each subarea or district allocated under the CDPs and the availability of the findings, will be published in the Federal Register.

(iv) NMFS will allocate no more than 33 percent of the total CDQ to any approved CDP application.

(v) A CDQ community may not concurrently receive more than one pollock, halibut, or sablefish allocation and only one application for each type of CDP per CDQ applicant will be accepted.

(2) Not consistent with criteria. (i) If NMFS finds that the Governor's recommendations for CDQ allocations are not consistent with the evaluation criteria set forth in these regulations and disapproves the Governor's recommendations, NMFS shall so advise the Governor and the Council in writing, including the reasons therefor.

(ii) Notification of the decision will be published in the Federal Register.

(3) Revised CDP. (i) The CDP applicant may submit a revised CDP to the Governor for submission to NMFS.

(ii) Review by NMFS of a revised CDP application will be in accordance with the provisions set forth in this section.

(d) Evaluation criteria. NMFS will approve the Governor's recommendations for CDPs if NMFS finds the CDP is consistent with the requirements of these regulations, including the following:

(1) CDP application. Each CDP application is submitted in compliance with the application procedures described in paragraph (b) of this section.

(2) NMFS review. Prior to approval of a CDP recommended by the Governor, NMFS will review the Governor's findings to determine that each community that is part of a CDP is listed in Table 7 of this part or meets the following criteria for an eligible community:

(i) The community is located within 50 nm from the baseline from which the breadth of the territorial sea is measured along the Bering Sea coast from the Bering Strait to the western most of the Aleutian Islands, or on an island within the Bering Sea. A community is not eligible if it is located on the GOA coast of the North Pacific Ocean, even if it is within 50 nm of the baseline of the Bering Sea.

(ii) The community is certified by the Secretary of the Interior pursuant to the Native Claims Settlement Act (Public Law 92-203) to be a native village.

(iii) The residents of the community conduct more than half of their current commercial or subsistence fishing effort in the waters of the BSAI.

(iv) The community has not previously developed harvesting or processing capability sufficient to support substantial groundfish fisheries participation in the BSAI, unless the community can show that benefits from an approved CDP would be the only way to realize a return from previous investments. The community of Unalaska is excluded under this provision.

(3) Qualified managing organization. Each CDP application demonstrates that a qualified managing organization will be responsible for the harvest and use of the CDQ allocation pursuant to the CDP.

(4) Exceeding the CDQ allocation. Each CDP application demonstrates that its managing organization can effectively prevent exceeding the CDQ allocation.

(5) Governor's findings. The Governor has found for each recommended CDP that:

(i) The CDP and the managing organization are fully described in the CDP application, and have the ability to successfully meet the CDP milestones and schedule.

(ii) The managing organization has an adequate budget for implementing the CDP, and the CDP is likely to be successful.

(iii) A qualified applicant has submitted the CDP application and the applicant and managing organization have the support of each community participating in the proposed CDQ project as demonstrated through an official letter approved by the governing body of each such community.

(iv) The following factors have been considered:

(A) The number of individuals from applicant communities who will be employed under the CDP, the nature of their work, and career advancement.

(B) The number and percentage of low income persons residing in the applicant communities, and the economic opportunities provided to them through employment under the CDP.

(C) The number of communities cooperating in the application.

(D) The relative benefits to be derived by participating communities and the specific plans for developing a self-sustaining fisheries economy.

(E) The success or failure of the applicant and/or the managing organization in the execution of a prior CDP (e.g., exceeding a CDQ allocation or any other related violation may be considered a failure and may therefore result in partially or fully precluding a CDP from a future CDQ allocation).

(6) Qualified applicant. For purposes of this paragraph (d), "qualified applicant" means:

(i) A local fishermen's organization from an eligible community, or group of eligible communities, that is incorporated under the laws of the State of Alaska, or under Federal law, and whose board of directors is composed of at least 75 percent resident fishermen of the community (or group of communities) that is (are) making an application; or

(ii) A local economic development organization incorporated under the laws of the State of Alaska, or under Federal law, specifically for the purpose of designing and implementing a CDP, and that has a board of directors composed of at least 75 percent resident fishermen of the community (or group of communities) that is (are) making an application.

(7) Resident fisherman. For the purpose of this paragraph (d), "resident fisherman" means an individual with documented commercial or subsistence fishing activity who maintains a mailing address and permanent domicile in the community and is eligible to receive an Alaska Permanent Fund dividend at that address.

(8) Board of directors. If a qualified applicant represents more than one community, the board of directors of the applicant must include at least one member from each of the communities represented.

(e) Monitoring of CDPs—(1) CDP reports. The following reports must be submitted to NMFS:

(i) Annual progress reports. (A) CDP applicants are required to submit annual progress reports to the Governor by June 30 of the year following allocation.

(B) Annual progress reports will include information describing how the CDP has met its milestones, goals, and objectives.

(C) On the basis of those reports, the Governor will submit an annual progress report to NMFS and recommend whether CDPs should be continued.

(D) NMFS must notify the Governor in writing within 45 days of receipt of the Governor's annual progress report, accepting or rejecting the annual progress report and the Governor's recommendations.

(E) If NMFS rejects the Governor's annual progress report, NMFS will return it for revision and resubmission.

(F) The report will be deemed approved if NMFS does not notify the Governor in writing within 45 days of the report's receipt.

(ii) Annual budget report. (A) An annual budget report is a detailed estimation of income and expenditures for each CDP project as described in paragraph (b)(1)(i) of this section for a calendar year.

(B) The annual budget report must be submitted to NMFS by December 15 preceding the year for which the annual budget applies.

(C) Annual budget reports are approved upon receipt by NMFS, unless disapproved in writing by December 31. If disapproved, the annual budget report may be revised and resubmitted to NMFS.

(D) NMFS will approve or disapprove a resubmitted annual budget report in writing.

(iii) Annual budget reconciliation report. A CDQ group must reconcile each annual budget by May 30 of the year following the year for which the annual budget applied. Reconciliation is an accounting of the annual budget's estimated income and expenditures with the actual income and expenditures, including the variance in dollars and variance in percentage for each CDP project that is described in paragraph
(b)(1)(i) of this section. If a general budget, as described in paragraph (b)(2)(vii) of this section, is no longer correct due to the reconciliation of an annual budget,

then the general budget must also be revised to reflect the annual budget reconciliation. The revised general budget must be included with the annual budget reconciliation report.

(2) Increase in CDQ allocation. If an applicant requests an increase in a CDQ, the applicant must submit a new CDP application for review by the Governor and approval by NMFS as described in paragraphs (b) and (c) of this section.

(3) Substantial amendments. (i) A CDP is a working business plan and must be kept up to date. Substantial amendments, as described in paragraph (e)(3)(iv) of this section, to a CDP will require written notification to the Governor and subsequent approval by the Governor and NMFS before any change in a CDP can occur. The Governor may recommend to NMFS that the request for an amendment be approved.

(ii) NMFS may notify the Governor in writing of approval or disapproval of the amendment within 30 days of receipt of the Governor's recommendation. The Governor's recommendation for approval of an amendment will be deemed approved if NMFS does not notify the Governor in writing within 30 calendar days of receipt of the Governor's recommendation.

(iii) If NMFS determines that the CDP, if changed, would no longer meet the criteria under paragraph (d) of this section, or if any of the requirements under this section would not be met, NMFS shall notify the Governor in writing of the reasons why the amendment cannot be approved.

(iv) For the purposes of this section, substantial amendments are defined as changes in a CDP, including, but not limited to, any of the following:

(A) Any change in the applicant communities or replacement of the managing organization.

(B) A change in the CDP applicant's harvesting or processing partner.

(C) Funding a CDP project in excess of $100,000 that is not part of an approved general budget.

(D) More than a 20-percent increase in the annual budget of an approved CDP project.

(E) More than a 20-percent increase in actual expenditures over the approved annual budget for administrative operations.

(F) A change in the contractual agreement(s) between the CDP applicant and its harvesting or processing partner, or a change in a CDP project, if such change is deemed by the Governor or NMFS to be a material change.

(v) Notification of an amendment to a CDP shall include the following information:

(A) The background and justification for the amendment that explains why the proposed amendment is necessary and appropriate.

(B) An explanation of why the proposed change to the CDP is an amendment according to paragraph (e)(3)(i) of this section.

(C) A description of the proposed amendment, explaining all changes to the CDP that result from the proposed amendment.

(D) A comparison of the original CDP text with the text of the proposed changes to the CDP, and the changed pages of the CDP for replacement in the CDP binder.

(E) Identification of any NMFS findings that would need to be modified if the amendment is approved along with the proposed modified text.

(F) A description of how the proposed amendment meets the requirements of this subpart. Only those CDQ regulations that are affected by the proposed amendment need to be discussed.

(4) Technical amendments. (i) Any change to a CDP that is not a substantial amendment as defined in paragraph (e)(3)(iv) of this section is a technical amendment. It is the responsibility of the CDQ group to coordinate with the Governor to ensure that a proposed technical amendment does not meet the definition for a substantial amendment. Technical amendments require written notification to the Governor and NMFS before the change in a CDP occurs.

(ii) A technical amendment will be approved when the CDQ group receives a written notification from NMFS announcing the receipt of the technical amendment. The Governor may recommend to NMFS, in writing, that a technical amendment be disapproved at any time. NMFS may disapprove a technical amendment in writing at any time, with the reasons therefore.

(iii) Notification should include:

(A) The pages of the CDP, with the text highlighted to show deletions and additions.

(B) The changed pages of the CDP for replacement in the CDP binder.

(5) Cease fishing operations. It is the responsibility of the CDQ-managing organization to cease fishing operations once a CDQ allocation has been reached.

(f) Suspension or termination of a CDP—(1) Governor's recommendation.

(i) NMFS, at any time, may partially suspend, suspend, or terminate any CDP upon written recommendation of the Governor setting out his or her reasons that the CDP recipient is not complying with these regulations.

(ii) After review of the Governor's recommendation and reasons for a partial suspension, suspension, or termination of a CDP, NMFS will notify the Governor in writing of approval or disapproval of his or her recommendation within 45 days of its receipt.

(iii) In the event of approval of the Governor's recommendation, NMFS will publish an announcement in the Federal Register that the CDP has been partially suspended, suspended, or terminated, along with reasons therefore.

(2) Non-compliance. NMFS also may partially suspend, suspend, or terminate any CDP at any time if NMFS finds a recipient of a CDQ allocation pursuant to the CDP is not complying with these regulations, other regulations, or provisions of the Magnuson Act or other applicable law. Publication of suspension or termination will appear in the Federal Register, along with the reasons therefor.

(3) Review of allocation. An annual progress report, required under paragraph (e)(1)(i) of this section, will be used by the Governor to review each CDP to determine whether the CDP and CDQ allocation thereunder should be continued, decreased, partially suspended, suspended, or terminated under the following circumstances:

(i) If the Governor determines that the CDP will successfully meet its goals and objectives, the CDP may continue without any Secretarial action.

(ii) If the Governor recommends to NMFS that an allocation be decreased, the Governor's recommendation for decrease will be deemed approved if NMFS does not notify the Governor, in writing, within 30 days of receipt of the Governor's recommendation.

(iii) If the Governor determines that a CDP has not successfully met its goals and objectives, or appears unlikely to become successful, the Governor may submit a recommendation to NMFS that the CDP be partially suspended, suspended, or terminated. The Governor must set out, in writing, his or her reasons for recommending suspension or termination of the CDP.

(iv) After review of the Governor's recommendation and reasons therefor, NMFS will notify the Governor, in writing, of approval or disapproval of his or her recommendation within 30 days of its receipt. In the case of suspension or termination, NMFS will publish notification in the Federal Register, with reasons therefor. [61 FR 31230, June 19, 1996, as amended at 61 FR 35579, July 5, 1996; 61 FR 41745, Aug. 12, 1996]

Code of Federal Regulations
Subpart C—Western Alaska Community Development Quota Program
Sec. 679.31 CDQ reserve.

(a) Pollock CDQ reserve (applicable through December 31, 1998). (1) In the proposed and final harvest specifications required under Sec. 679.20(c), one-half of the pollock TAC placed in the reserve for each subarea or district will be assigned to a CDQ reserve for each subarea or district.

(2) NMFS may add any amount of a CDQ reserve back to the nonspecific reserve if, after September 30, the Regional Director determines that amount will not be used during the remainder of the fishing year.

(b) Halibut CDQ reserve. (1) NMFS will annually withhold from IFQ allocation the proportions of the halibut catch limit that are specified in this paragraph (b) for use as a CDQ reserve.

(2) Portions of the CDQ for each specified IPHC regulatory area may be allocated for the exclusive use of an eligible Western Alaska community or group of communities in accordance with a CDP approved by the Governor in consultation with the Council and approved by NMFS.

(3) The proportions of the halibut catch limit annually withheld for purposes of the CDQ program, exclusive of issued QS, are as follows for each IPHC regulatory area:

(i) Area 4B. In IPHC regulatory area 4B, 20 percent of the annual halibut quota shall be made available for the halibut CDQ program to eligible communities physically located in or proximate to this regulatory area. For the purposes of this section, "proximate to" an IPHC regulatory area means within 10 nm from the point where the boundary of the IPHC regulatory area intersects land.

(ii) Area 4C. In IPHC regulatory area 4C, 50 percent of the halibut quota shall be made available for the halibut CDQ program to eligible communities physically located in IPHC regulatory area 4C.

(iii) Area 4D. In IPHC regulatory area 4D, 30 percent of the halibut quota shall be made available for the halibut CDQ program to eligible communities located in or proximate to IPHC regulatory areas 4D and 4E.

(iv) Area 4E. In IPHC regulatory area 4E, 100 percent of the halibut quota shall be made available for the halibut CDQ program to communities located in or proximate to IPHC regulatory area 4E. A fishing trip limit of 6,000 lb (2.7 mt) will apply to halibut CDQ harvesting in IPHC regulatory area 4E.

(c) Sablefish CDQ reserve. In the proposed and final harvest limit specifications required under Sec. 679.20(c), NMFS will specify 20 percent of the fixed gear allocation of sablefish in each BSAI subarea as a sablefish CDQ reserve, exclusive of issued QS. Portions of the CDQ reserve for each subarea may be allocated for the exclusive use of CDQ applicants in accordance with CDPs approved by the Governor in consultation with the Council and approved by NMFS. NMFS will allocate no more than 33 percent of the total CDQ for all subareas combined to any one applicant with an approved CDP application.

Code of Federal Regulations
Subpart C—Western Alaska Community Development Quota Program
Sec. 679.32 Estimation of total pollock harvest in the CDQ fisheries (applicable through December 31, 1998).

(a) Recordkeeping and reporting requirements. Vessels and processors participating in pollock CDQ fisheries must comply with recordkeeping and reporting requirements set out at Sec. 679.5.

(b) Total pollock harvests—(1) Observer estimates. Total pollock harvests for each CDP will be determined by observer estimates of total catch and catch composition, as reported on the daily observer catch message.

(2) Cease fishing. The CDQ-managing organization must arrange to receive a copy of the observer daily catch message from processors in a manner that allows the CDQ-managing organization to inform processors to cease fishing operations before the CDQ allocation has been exceeded. CDQ-managing organization representatives must also inform NMFS within 24 hours after the CDQ has been reached and fishing has ceased.

(3) NMFS estimates. If NMFS determines that the observer, the processor, or the CDQ-managing organization failed to follow the procedures described in paragraphs (c), (d), and (e) of this section for estimating the total harvest of pollock, or violated any other regulation in this subpart C of this part, NMFS reserves the right to estimate the total pollock harvest based on the best available data.

(c) Observer coverage. Vessel operators and processors participating in CDQ fisheries must comply with the following requirements for observer coverage:

(1) Shoreside processor. (i) Each shoreside processor participating in the CDQ fisheries must have one NMFS-certified observer present at all times while groundfish harvested under a CDQ are being received or processed.

(ii) The Regional Director is authorized to require more than one observer for a shoreside processor if:

(A) The CDQ delivery schedule requires an observer to be on duty more than 12 hours in a 24-hour period;

(B) Simultaneous deliveries of CDQ harvests by more than one vessel cannot be monitored by a single observer; or

(C) One observer is not capable of adequately monitoring CDQ deliveries.

(2) Processor vessel. Each processor vessel participating in the CDQ fisheries must have two NMFS-certified observers aboard the vessel at all times while groundfish harvested under a CDQ are being harvested, processed, or received from another vessel.

(3) Catcher vessel. Observer coverage requirements for catcher vessels participating in the CDQ fisheries are in addition to any observer coverage requirements in subpart E of this part. Each catcher vessel delivering groundfish harvested under a CDQ, other than a catcher vessel delivering only unsorted codends to a processor or another vessel, must have a NMFS-certified observer on the vessel at all times while the vessel is participating in the CDQ fisheries, regardless of the vessel length.

(d) Shoreside processor equipment and operational requirements. Each shoreside processor participating in the CDQ fisheries must comply with the following requirements:

(1) Certified scale. Groundfish harvested in the CDQ fisheries must be recorded

and weighed on a scale certified by the State of Alaska. Such a scale must measure catch weights at all times to at least 95-percent accuracy, as determined by a NMFS-certified observer or authorized officer. The scale and scale display must be visible simultaneously by the observer.

(2) Access to scale. Observers must be provided access to the scale used to weigh groundfish landings.

(3) Retention of scale printouts. Printouts of scale measurements of each CDQ delivery must be made available to observers and be maintained in the shoreside processor for the duration of the fishing year, or for as long after a fishing year as product from fish harvested during that year are retained in the shoreside processor.

(4) Prior notice of offloading schedule. The manager of each shoreside processor must notify the observer(s) of the offloading schedule of each CDQ groundfish delivery at least 1 hour prior to offloading to provide the observer an opportunity to monitor the weighing of the entire delivery.

(e) Processor vessel measurement requirements. Each processor vessel participating in the CDQ fishery for pollock must estimate the total weight of its groundfish catch by the volumetric procedures specified in this paragraph (e).

(1) Receiving bins. Each processor vessel estimating its catch by volumetric measurement must have one or more receiving bins in which all fish catches are placed to determine total catch weight prior to sorting operations.

(2) Bin volume. The volume of each bin must be accurately measured, and the bin must be permanently marked and numbered in 10-cm increments on all internal sides of the bin. Marked increments, except those on the wall containing the viewing port or window, must be readable from the outside of the bin at all times. Bins must be lighted in a manner that allows marked increments to be read from the outside of the bin by a NMFS-certified observer or authorized officer.

(3) Bin certification. (i) The bin volume and marked and numbered increments must be certified by a registered engineer with no financial interest in fishing, fish processing, or fish tender vessels, or by a qualified organization that has been designated by the USCG Commandant, or an authorized representative thereof, for the purpose of classing or examining commercial fishing industry vessels under the provisions of 46 CFR 28.76.

(ii) Bin volumes and marked and numbered increments must be recertified each time a bin is structurally or physically changed.

(iii) The location of bin markings, as certified, must be described in writing. Tables certified under this paragraph (e)(1)(iii) indicating the volume of each certified bin in cubic meters for each 10-cm increment marked on the sides of the bins, must be submitted to the NMFS Observer Program prior to harvesting or receiving groundfish and must be maintained on board the vessel and made available to NMFS-certified observers at all times.

(iv) All bin certification documents must be dated and signed by the certifier.

(4) Prior notification. Vessel operators must notify observers prior to any removal or addition of fish from each bin used for volumetric measurements of catch in such a manner that allows an observer to take bin volume measurements prior to fish being removed from or added to the bin. Once a volumetric measurement has been taken, additional fish may not be added to the bin until at least half the original volume has been removed. Fish may not be removed from or added to a bin used for volumetric measurements of catch until an observer indicates that bin volume measurements have been completed and any samples of catch required by the observer have been taken.

(5) Separation of fish. Fish from separate hauls or deliveries from separate harvesting vessels may not be mixed in any bin used for volumetric measurements of catch.

(6) Bin viewing port. The receiving bins must not be filled in a manner that obstructs the viewing ports or prevents the observer from seeing the level of fish throughout the bin. [61 FR 31230, June 19, 1996; 61 FR 37843, July 22, 1996 as amended at 61 FR 41745, Aug. 12, 1996]

Code of Federal Regulations
Subpart C—Western Alaska Community Development Quota Program
Sec. 679.33 Halibut and sablefish CDQ.

(a) Permits. The Regional Director will issue a halibut and/or sablefish CDQ permit to the managing organization responsible for carrying out an approved CDQ project. A copy of the halibut and/or sablefish CDQ permit must be carried on any fishing vessel operated by or for the managing organization, and be made available for inspection by an authorized officer. Each halibut and/or sablefish CDQ permit will be non-transferable and will be effective for the duration of the CDQ project or until revoked, suspended, or modified.

(b) CDQ cards. The Regional Director will issue halibut and/or sablefish CDQ cards to all individuals named on an approved CDP application. Each halibut and/or sablefish CDQ card will identify a CDQ permit number and the individual authorized by the managing organization to land halibut and/or sablefish for debit against its CDQ allocation.

(c) Alteration. No person may alter, erase, or mutilate a halibut and/or sablefish CDQ permit, card, registered buyer permit, or any valid and current permit or document issued under this part. Any such permit, card, or document that has been intentionally altered, erased, or mutilated will be invalid.

(d) Landings. All landings of halibut and/or sablefish harvested under an approved CDQ project, dockside sales, and outside landings of halibut and/or sablefish must be landed by a person with a valid halibut and/or sablefish CDQ card to a person with a valid registered buyer permit, and reported in compliance with Sec. 679.5 (l)(1) and (l)(2).

(e) CDQ fishing seasons. See Sec. 679.23(e)(4).

Code of Federal Regulations
Subpart C—Western Alaska Community Development Quota Program
Sec. 679.34 CDQ halibut and sablefish determinations and appeals.

Section 679.34 describes the procedure for appealing initial administrative determinations for the halibut and sablefish CDQ program made under this subpart C of this part.

Appendix F

Investments Pursued by CDQ Groups

**ALEUTIAN PRIBILOF ISLAND COMMUNITY
DEVELOPMENT ASSOCIATION (APICDA)**

Vessel Acquisitions

Vessel	% of Ownership	Description
F/V Golden Dawn	25%	The Golden Dawn is a 148 foot pollock catcher vessel operated by Trident Seafoods.
F/V Ocean Prowler	25%	The Ocean Prowler is a 155 foot longline processing vessel.
F/V Prowler	25%	The Prowler is a 115 foot longline processing vessel.
F/V Stardust	100%	The Stardust is a versatile 58 foot longline/crab vessel.
F/V Bonanza	100%	The Bonanza is a versatile 58 foot longline/crab vessel.

Vessel	% of Ownership	Description
AP#1 AP#2 AP#3	100%	Built in 1994, APICDA has three 32-foot longline vessels that operate out of Atka in the halibut fishery. APICDA has another 26foot longline vessel in Atka. (Vessel name not provided).
Grand Aleutian	100%	The Grand Aleutian is a 32 foot sport fishing charter vessel working out of Dutch Harbor.
F/V Rebecca B (Destroyed 1996)	40%	The Rebecca B was a longline processing vessel owned in a partnership with YDFDA. The vessel ran aground in 1996 and was destroyed.

Community Based Fisheries Development Projects

Project Title	Community	Description
St. George Dredging	St. George	In 1993, APICDA contributed to a project to dredge Zapadni Bay Harbor.
False Pass Dock Improvements	False Pass	In 1993, APICDA contributed funds for the extension of water and sewer services to the False Pass dock.
False Pass Gear Storage	False Pass	Starting in 1993, APICDA has constructed a gear storage facility in False Pass intended to service salmon fishermen in the area.
Atka Pride Seafoods	Atka	In a 50/50% joint venture with Atka Fishermen's Association, APICDA has provided vital capital to renovate a halibut processing plant.
Nelson Lagoon	Nelson Lagoon	In 1995, APICDA constructed a dock in Dock Nelson Lagoon.
False Pass Harbor Improvements	False Pass	Assisted in funding a boat launch ramp in 1996.

Project Title	Community	Description
Kaynx, Inc.	St. George	Starting in 1996, Kayux, a 50/50% joint venture between the APICDA and St. George Tanaq Corporation (local native corporation), is a harbor development project intended to attract seafood processors to the area.
Atka Dock Facility	Atka	Built in 1997 in concert with federal and state funds, APICDA has coordinated the construction of a large dock with requisite facilities and a transient camp, located adjacent to Atka. APICDA is trying to make Atka a commercial center for the Bering Sea fishery.
Nelson Lagoon Gear Storage	Nelson Lagoon	Starting in 1997, APICDA has constructed a gear storage facility in Nelson Lagoon intended to service salmon fishermen in the area.
Processing facilities	Various	APICDA plans to research and evaluate the possibility of processing facilities in several of its communities. If the evaluations shows the plants to be profitable, APICDA will begin development.

Other Fisheries Development Projects

Project Title	Description
Product Diversification Program	In partnership with Trident/Starbound, this program looks to develop new products with pollock and other CDQ species.
IFQ Fund	APICDA puts aside funds for loans to residents who want to purchase IFQs.

Project Title	Description
IFQ Purchases	APICDA has purchased several thousand shares of halibut and sablefish IFQs.
Ocean Logic, L.L.C.	In partnership with YDFDA, Ocean Logic is a software development project intended for use aboard fishing vessels in order to track and manage harvest data.

Employment Opportunities

Trident/Starbound offers a preferential hire program for qualified residents of APICDA's region. They also provide training when needed and are investigating the establishment of a shoreside training program. Opportunities vary from clerical and processing jobs to maintenance and equipment operators.

Other Fishing Employment

A great deal of APICDA's reported employment in this category comes from the activities that have occurred in Atka with the development of the processing facility and increase in harvesting capacity. Other fisheries related work has come from employment opportunities made available through sablefish arrangements APICDA has with larger vessels.

Vocational Training

APICDA offers scholarships up to $3,000 for vocational training in occupations that support commercial fisheries or community development. Programs have included diesel mechanics, longshoring, office skills, and small business administration.

On-the-job training is available for individuals who have completed vocational program. Positions are with Trident and Starbound Partnership to help vocational graduates find permanent job placements.

Scholarships

High school graduates and college students are eligible for college scholarships. Scholarship amounts vary and are determined annually by the APICDA board of directors. APICDA also publishes a booklet outlining other academic scholarships available to students.

Internships

Pacific Associates sponsors an internship in the Juneau APICDA office in business development and management.

Trident Seafoods Corporation and the Starbound Partnership are developing an internship for outstanding applicants interested in careers in the seafood industry. Interns concentrate on specific interest areas such as product marketing, plant operation, corporate management and sales.

BRISTOL BAY ECONOMIC DEVELOPMENT CORPORATION (BBEDC)

Vessel Acquisitions

Vessel	% of Ownership	Description
F/V Arctic Fjord	20%	The Arctic Fjord is a 270-foot factory trawler. Managed by partner Arctic Storm, the vessel harvests pollock and other groundfish.
F/V Bristol Leader	50%	The Bristol Leader is a 167-foot freezer longliner. It is co-owned by Alaskan Leader Fisheries. It will harvest cod, halibut and sablefish.
F/V Bristol Mariner	45%	The Bristol Mariner is a 125 crab vessel. It is co-owned by Kaldestad Fisheries.
F/V Nordic Mariner	45%	The Bristol Mariner is a 121-foot crab harvesting vessel. It is co-owned by Kaldestad Fisheries.

Community Based Fisheries Development Projects

Project Title	Community	Description
Regional Business Development	Available to all	BBEDC will set aside funds to assist in testing a project's feasibility and potential implementation.

Project Title	Community	Description
Inshore Halibut Fishery	Available to all	BBEDC provided harvest management services for the region's CDQ halibut fisheries.
Regional Infrastructure Development	Available to all	BBEDC will set aside of funds for a regionally coordinated effort to add to Bristol Bay's fishing infrastructure.

Other Fisheries Development Projects

Project Title	Description
Permit Retention and Brokerage	Since 1993, BBEDC has devoted staff end funds to assist region fishermen in keeping or acquiring their fishing permits. More information provided below.
IFQ Purchases	BBEDC has made a substantial investment in sablefish halibut IFQ shares for the Bering Sea and Gulf of Alaska. These IFQ shares will supplement the investment in the F/V Bristol Leader.
Alaska Seafood Investment	BBEDC has established the Alaska Seafood Investment Fund (ASIF) to make investments in Alaskan seafood businesses. These investments will be made outside of Bristol Bay's fully developed sockeye salmon and fisheries.
Regional Fisheries Development Project	This project identifies and tests the feasibility of fisheries related economic projects. BBEDC will have an ongoing agenda item to thoroughly survey the Bristol Bay waters to test the feasibility of new fisheries.
Sea State	In coordination with 3 other CDQ groups, BBEDC is developing a real time data tracking and catch accounting system to use during the CDQ fishery.

Employment Opportunities

Employment on all Arctic Storm, Inc vessels (F/T Arctic Storm and F/T Arctic Fjord) is open to all Bristol Bay residents. However, priority is given to residents of BBEDC member communities. There are two main seasons for employment with Arctic Storm, Inc. The "A" season which generally is from January to March and the "B" season which is generally from September to November.

The critical times to apply for these jobs are November and December for "A" season and June and July for "B" season. New hires start out as processing technicians and may work into positions of greater responsibility.

Applications for work on long-line and crab catcher and processing vessels are accepted year round. They fish at various times of the year and there is usually a short notice for job openings. All applications are kept on file and when there is a job available the qualified applicants are notified in the order that they were submitted.

BBEDC has also arranged for employment opportunities with our crab-processing partner. Work is available at various times throughout the year in a number of locations in Alaska. These jobs are on both floating processors and shore plants.

Contact Program Manager James Sifsof at BBEDC the headquarters in Dillingham to inquire about any of these employment opportunities.

Vocational Training

BBEDC offers opportunities for vocational training and provides funds to supplement other vocational and technical grants within the region. Advanced vocational and technical training has included office occupations, food service, electronics repair and diesel mechanics, and other seafood related fields of study. Contact Program Manager James Sifsof at BBEDC the headquarters in Dillingham for course offerings, information, and applications.

The Voc-Tech Program also provides financial assistance to people who have previously worked on our vessels and have a desire to upgrade their training and skills in order to qualify for advanced positions on board these vessels.

In addition, BBEDC has an agreement with the University of Alaska in Dillingham to provide Adult Basic Education (ABE) and Grade Equivalent Diploma (GED) classes in member communities. Contact Kim Fortune, University of Alaska, at 1-800-478-5109 or 842-5109 for class schedules and information.

Scholarships

The Harvey Samuelson Scholarship Trust has been set up in a similar manner to the Alaska Permanent Fund. The interest earned from the trust fund is used to provide scholarship awards to Bristol Bay residents. This will enable the trust to stand alone and be available for Bristol Bay residents' academic needs in the foreseeable future.

BBEDC offers academic scholarships to college students taking undergraduate and graduate courses. The program is open to residents of BBEDC member communities. Scholarship applications are due by July 31.

For more information contact Program Manager James Sifsof at BBEDC the headquarters or Pearl Strub, BBNA Higher Education Coordinator, at 1-800-478-5257.

Internships

BBEDC offers a variety of internships for qualified Bristol Bay residents. Interns may gain professional experience in our harvesting partner's corporate offices in Seattle and Kodiak. There are also opportunities to gain work experience with the Alaska Department of Fish and Game. High school students are eligible for clerical internships in the BBEDC office each quarter during the school year.

Contact Program Manager James Sifsof at BBEDC the headquarters in Dillingham for more information.

Bristol Bay Permit Brokerage

This program meets a very important and continuing need in Bristol Bay. The program was developed to help maintain regional ownership of Bristol Bay limited entry salmon permits.

The Bristol Bay Permit Brokerage (BBPB) assists Bristol Bay residents in buying, selling, and transferring commercial fishing permits and arranging boat leases. The Brokerage also assists in the leasing and selling of fishing vessels. In addition, financial counseling is available for residents needing advice on meeting IRS tax obligations, boat and permit loans, and child support.

CENTRAL BERING SEA FISHERMEN'S ASSOCIATION (CBSFA)

Vessel Acquisitions

Vessel	% of Ownership	Description
F/V Zolotoi	20%	Purchased in 1994, the Zolotoi is a 98 foot crab vessel.
Longline Vessel	Undetermined	Through the expanded species program, CBSFA intents to invest in a longline vessel.
Crab Harvesting Vessel	Undetermined	Through the expanded species program, CBSFA intents to invest in a longline vessel.

Community Based Fisheries Development Projects

Project Title	Community	Description
Harbor Dredging	St. Paul	Starting in 1994, CBSFA performed dredging activities in the harbor area.
Small Dock Moorage	St. Paul	CBSFA has maintained and funded a small dock used to moor the small vessel fleet during the halibut fishery.
Harbor Development	St. Paul	CBSFA is participating in the harbor development project in coordination with the Army Corps of Engineers, TDX (local native corporation) and the City of St. Paul.
Small Boat Harbor	St. Paul	As a component of the larger harbor, CBSFA is planning to build a permanent boat harbor for its halibut fleet.

Other Fisheries Development Projects

Project Title	Description
Impact Fund	CBSFA has set aside a small impact fund used to support cultural activities.
Revolving Loan Program	Since 1993, CBSFA has operated a revolving loan program established to provide boat and gear loans to resident fishermen.
Sea State	In coordination with 3 other CDQ groups, CBSFA is developing a real time data tracking and catch accounting system to use during the CDQ fishery.

Employment Opportunities

St. Paul Community members, TDX Corporation shareholders, and CBFSA members are eligible for employment with the American Seafoods Company and Pribilof Bering Seafood, Ltd. In addition, there are also opportunities with Jubilee Seafoods and onboard our crab harvesting partners' vessels, F/V Scandies Rose, F/V Zolotoi and F/V Kona Kai. New hires generally start out as processors but may work into positions of greater responsibility.

Contact Carl Merculief or Kathy Faltz at the CBSFA headquarters in Anchorage, at (907) 279-6566, for more information. Details of a special "hands-on" training being offered by one of our crab harvesters will be available soon.

Vocational Training

CBFSA will provide tuition for vocational training programs relating to the fishing industry and for occupations that improve and support community development.

Training programs have included areas such as computer skills, small business administration, heavy equipment operation, welding, and automotive repair.

Scholarships

CBFSA funds a scholarship program through a grant from American Seafoods. Full time college students maintaining a 2.0 grade point average may receive a $2,750 scholarship each semester.

Internships

CBFSA sponsors an internship program in conjunction with the American Seafoods Company.The intern works in the American Seafoods corporate headquarters in Seattle and gains experience in corporate management and operations.

Loan Program

CBFSA provides loans to qualifying applicants for the purchase of commercial vessels and gear. Contact Phillip Lestenkof in St. Paul at 546-2579 and Kathy Faltz, Administration Manager in Anchorage office at (907) 279-6566. CBFSA also has an IFQ Loan Program for qualified applicants to purchase halibut IFQs in Area 4C.

COASTAL VILLAGES REGIONAL FUND (CVRF)

Vessel Acquisitions

Vessel	% of Ownership	Description
F/V Ocean Harvester	45%	The Ocean Harvester is a 58-foot longline vessel.
Crab Vessel	Undetermined	Through the expanded species program, CVRF intends to invest in a crab vessel.
Head & Gut Vessel	Undetermined	Through the expanded species program, CVRF intends to invest in a head and gut vessel.

Community Based Fisheries Development Projects

Project Title	Community	Description
Funding of halibut processing plants	Toksook Bay, Tununak, Mekoryuk, Chevak	Since 1994, CVRF has provided small loans for working capital to halibut processing facilities.
Kuskokwim processing facility	Bethel	If it proves feasible, CVRF will consider another salmon processing operation near Bethel.

Project Title	Community	Description
Quinhagak salmon processing plant	Quinhagak	If it proves feasible, CVRF will consider funding renovations to a Quinhagak salmon processing plant.

Other Fisheries Development Projects

Project Title	Description
Coastal Village Investment Fund	With tax set asides from the IP, CVIF is intended to provide capital for new economic activity in the region.
IFQ Assistance	CVRF has assisted region residents in securing IFQs.
Revolving Loan Fund	In conjunction with YDFDA and Alaska Village Council Presidents, CVRF contributes funds for boat and gear loans.
Sea State	In coordination with 3 other CDQ groups, CVRF is developing a real time data tracking and catch accounting system to use during the CDQ fishery.
4-SITE Program	This program is a comprehensive training and employment program that seeks to address the qualities of residents when placing them in a position for career advancement.
Salmon Roe University	CVRF contracts with Sheldon Jackson University to provide residents with roe technician instruction.
Tax and Permit Assistance Program	CVRF works to preserve fishing permits in the region.

Employment Opportunities

Residents of CVRF member communities and non-member villages throughout the Yukon-Kuskokwim region are eligible for employment with a variety of seafood companies both on shore and at sea.

New hires generally start out in entry level processing positions with opportunities to work into positions of greater responsibilities. Applicants must have a physical examinations and a drug screening test to qualify.

Vocational Training

CVRF has established a seafood employment-training program in conjunction with Sheldon Jackson College in Sitka. The program includes HACCP training and prepares participants for positions as salmon, herring, and pollock roe technicians.

CVRF also sponsors technical training for specialized areas such as fish processing machine maintenance and aluminum welding.

Scholarships

The Louis Bunyan Memorial Scholarship program offers academic and vocational scholorships to eligible high school graduates and college students from CVRF member communities and nonmember villages in the Yukon-Kuskokwim region.

Internships

Internships at are available to high school graduates and college student pursuing careers related to commercial fisheries.

Apprenticeships are available at the Westward Seafood Plant in Dutch Harbor for office and factory positions. Interns and apprentices may gain experience in plant operations, production, corporate management, human resources, and accounting.

NORTON SOUND ECONOMIC DEVELOPMENT CORPORATION (NSEDC)

Vessel Acquisitions

Vessel	% of Ownership	Description
Glacier Fish Company	50%	Glacier Fish Company is a fishing company with factory trawlers, the 201-foot Northern Glacier and the 276-foot Pacific Glacier, the FN Norton Sound and a seafood marketing arm. (See below)
F/V Norton Sound	49%	Owned jointly with GFC as the Norton Sound Fish Company, the Norton Sound is a 139-foot longline vessel with processing capability.
Golovin Bay Norton Bay	100% 100%	NSEDC purchased these two tender vessels and manages them under Norton Sound Vessel Management. The vessels, specially built for the Norton Sound region, will lower costs for NSSP and may provide another alternative for freight transportation.

Community Based Fisheries Development Projects

Project Title	Community	Description
Norton Sound Seafood Products	Various	NSSP is a for profit subsidiary that buys and markets salmon, crab, herring and halibut.
Koyuk Ice Machine	Koyuk	In 1993, NSEDC provided funds to assist in the purchase of an ice machine.
Unalakleet Processing Plant	Unalakleet	Since 1993, NSEDC has been assisting Unalakleet by providing funds for plant renovations and loans for reconstruction of the facility.

Project Title	Community	Description
Norton Sound Crab Company	Nome	NSCC is a fully owned subsidiary started in 1993 intended to serve as a crab processing facility.
Nome Floating Dock	Nome	In 1994, NSEDC provided matching funds for the construction of a floating dock in Nome.
Nome Eskimo Freezer Facility	Nome	In 1994, NSEDC funded improvements to the Nome Eskimo Freezer Facility.
Shaktoolik Facility Improvements	Shaktoolik	Since 1993, NSEDC has provided capital for a buying station in Shaktoolik.
Savoonga Halibut Improvements	Savoonga	To support the small halibut fishery that has been developed on Savoonga through the halibut CDQ, NSEDC has been funding renovations and additions to the fishery infrastructure.
St. Lawrence Island Halibut Fishery	St. Lawrence	In 1993, NSEDC established a commercial halibut fishery at St. Lawrence Island. This work included successful efforts to change International Pacific Halibut Commission (IPHC) regulations to establish an experimental fishery in Area 4D.
Nome Harbor Project	Nome	NSEDC may commit funds to a harbor development project in Nome.

Other Fisheries Development Projects

Project Title   Description

Salmon Enhancement Program   NSEDC has supported the formation of an
                                                     aquaculture association located in Elim.
                                                     The association will work to rebuild the
                                                     dwindling stocks of salmon vital to
                                                     subsistence activities in Norton Sound.

Revolving Loan Program   Since 1993, NSEDC has operated a
                                                     revolving loan program established to
                                                     provide permit, boat, and gear loans to
                                                     resident fishermen.

Salmon and Herring Marketing   NSEDC has organized salmon and herring
Program   buying/processing operations and will
                                                     conduct additional market research for
                                                     various products from the Norton Sound
                                                     fisheries.

Sea State   In coordination with 3 other CDQ groups,
                                                     NSEDC is developing real time data
                                                     tracking and catch accounting system
                                                     to use during the CDQ fishery.

Employment Opportunities

NSEDC offers employment opportunities for residents of member communities, including clerical and professional positions, in its local offices. The Unalakleet and Shaktoolik processing plants operate during the summer and hire processing workers, plant foreman, maintenance crews and equipment operators.

People interested in jobs on Glacier Fish Company vessels are required to complete basic seafood training program through the Alaska Vocational Technical Center (AVTEC). Applicants must have also a physical examination, pass a drug screening urinalysis and a complete hearing test.

Vocational Training

NSEDC sponsors a ten-day training program for basic seafood processing through AVTEC in Seward. Students who complete the program are eligible for employment with the Glacier Fish Company.

Scholarships

NSEDC awards annual scholarships of $1,000 to high school graduates or college students from member communities. Students wishing to attend an accredited college or vocational school may apply.

Internships

An internship is available for an outstanding applicant with the Glacier Fish Company. The intern works in the Glacier Fish Company office in Seattle and gains experience in management, plant operations, seafood production and marketing.

Loan Program

The NSEDC Revolving Loan Program makes loans up to $7,500 to purchase Norton Sound commercial fisheries entry permits. Loans of up to $15,000 are available for vessel improvements and up to $16,000 to participate in the summer crab fishery.

YUKON DELTA FISHERIES DEVELOPMENT ASSOCIATION (YDFDA)

Vessel Acquisitions

Vessel	% of Ownership	Description
F/V Blue Dolphin	100%	Purchased in 1993, the Blue Dolphin is a 47 foot longline/crab vessel.
F/V Nakat	100%	Purchased in 1993 and sold in 1997, the Nakat was a 53 foot longline/crab vessel.
Small Boat Fleet	100%	YDFDA owns and operates nine (9) 32 foot longline vessels.
F/V Lisa Marie	100%	Purchased in 1997, the Lisa Marie is a 78 foot trawl, pot and longline vessel.

Community Based Fisheries Development Projects

Project Title	Community	Description
Emmonak Value Added Processing Plant	Emmonak	Since 1993, YDFDA has provided loan funds to the Yukon Delta Fish Marketing Co-op for the purpose of plant construction and improvements.
Yukon Delta Fisheries	All	As a component of its training program, YDFDA operates YDF, which allows residents to fish the Bering Sea on small longline vessels.

Other Fisheries Development Projects

Project Title	Description
Salmon and Herring Permit Buy-Back Program	YDFDA has set up a permit buy back program in an attempt to retain fishing rights in the region.
Revolving Loan Fund	In conjunction with CVRF and Alaska Village Council Presidents YDFDA contributes funds for boat and gear loans.
Ocean Logic, L.L.C.	In partnership with APICDA, Ocean Logic is a software development project intended for use aboard fishing vessels in order to track and manage harvest data.
Exploratory Fishing Research	This program conducts research on the distribution, appropriate gear, and preferred fishing methods suitable for community based commercial fishing in the eastern Bering Sea.

Employment Opportunities

YDFDA's employment objectives are to provide on-the-job training and experience in offshore fisheries to community residents and provide immediate employment and income-earning opportunities to these residents. Although the pollock related employment opportunities with Golden Alaska have been some of the more lucrative found in the CDQ program, YDFDA has not stopped its employment recruitment efforts there. YDFDA continues to seek out other pollock companies to find employment for its residents.

The following is a list of pollock and non-pollock companies that YDFDA has worked with to provide employment for its region: American Seafoods, Westward Seafoods, Trident Seafoods, Starbound, O'Hara Corporation, Kodiak Fish Company, Fishermen's Finest, Peter Pan Seafoods, Premier Pacific, Supreme Alaska, Fanning Fisheries, and Seven Seas.

YDFDA provides substantial employment opportunities through its small boat fleet. The small boats are versatile in meeting the needs of several fisheries and continue to provide region residents with opportunities to sharpen their fishing skills.

YDFDA may be credited with assisting some residents in achieving future employment through their training efforts. YDFDA has assisted a few residents in starting welding businesses in their own communities after they completed training.

Vocational Training

YDFDA has created a unique training platform that folds actual fishing into a training setting. Through the vocational training offered at the Alaska Vocational Technical Center in Seward, region residents may choose any number of training courses that lend themselves to careers in the fishing industry. From there, YDFDA has limited space for residents to continue their education with on-the-job training aboard the small vessels.

Internships

YDFDA sponsors an internship in its Seattle office. Interns are mentored by YDFDA staff and gain experience in finance, management, and human resources.

Golden Alaska Seafoods offers a clerical internship in its Seattle headquarters for individuals interested in office occupations and operations.

Boat Loans and Leases

YDFDA has a revolving loan fund to finance Yukon River salmon and Norton Sound herring fishery permits. Applicants must apply through the YDFDA Seattle office and be approved by the YDFDA board of directors.

Graduates of the AVTEC longlining program may lease a 32-foot combination vessel. Boats are for longlining, pot fishing, jigging, and herring and gillnetting. Skippers awarded a lease may also purchase the boat from YDFDA.

Appendix G

Glossary

AAC (Alaska Administrative Code): Code of laws for the State of Alaska.

ADF&G (Alaska Department of Fish and Game): State agency responsible for overseeing the management of fish and wildlife within state jurisdiction.

Alia: Samoan fishing catamaran made of aluminum, or fiberglass and wood, approximately 30-feet long. Alia are used in various fisheries including trolling, longline, and bottomfishing.

American Samoa: An unincorporated territory of the United States in the southern Pacific Ocean South of Hawaii. American Samoa includes the islands of Manua, Tutuila, and the Rose and Swains Atolls.

ANCSA (Alaska Native Claims Settlement Act): Legislation passed in 1971 that provided monetary and land compensation for Alaskan natives as an agreement to relinquish additional land claims. The act established village corporations and regional corporations designed to conduct business operations from the initial settlement on behalf of native shareholders.

APICDA (Aleutian Pribilof Island Community Development Association): Community Development Quota Association composed of the villages of Akutan, Atka, False Pass, Nelson Lagoon, Nikolski, and St. George.

BBEDC (Bristol Bay Economic Development Corporation): Community Development Quota Association composed of the villages of Aleknagik, Clark's Point, Dillingham, Egegik, Ekuk, Manokotak, Naknek, King Salmon/Savonoski, South Naknek, Togiak, Twin Hills, Pilot Point/Ugashik, and Port Heiden.

biomass: The amount, or mass, of fish.

Board of Fisheries: Commission in the State of Alaska formed for purposes of the conservation and development of the fishery resources. The Board of Fisheries is composed of seven members appointed by the governor. The governor appoints each member on the basis of interest in public affairs, good judgment, knowledge, and ability in the field of action of the board, and with a view to providing diversity of interest and points of view in the membership. The appointed members are residents of the state and are appointed without regard to political affiliation or geographical location of residence. The commissioner is not a member of the Board of Fisheries. Board members serve three year terms.

BSAI (Bering Sea/Aleutian Islands): Area of the EEZ off the coast of Alaska including Bering Sea and the western side of the Aleutian Island chain.

bycatch: Fish caught in a fishery but discarded or released for economic or regulatory reasons. This does not include fish caught in a recreational catch and release program.

Carolinians: Natives of the Northern Mariana Islands.

catcher: Vessel that harvests fish but does not have on-board processing capacity.

catcher-processor: Vessel that can both catch and process the catch on-board. Also referred to as factory-trawlers in the North Pacific.

CBSFA (Central Bering Sea Fishermen's Association): Community Development Quota Association composed of the village of St. Paul.

CDP (Community Development Plan): Plan submitted to the State of Alaska by the CDQ association detailing the means of harvesting the quota allocation and proposing how the funds generated from the harvest or leasing of quota will be used. The CDP is required to detail the ways in which funds will be distributed and used in the CDQ program.

CDQ (Community Development Quota): A program in western Alaska under which a percentage of the total allowable catch of Bering Sea commercial fisheries are allocated to specific villages. The villages that are eligible for this program must be located within 50 miles of the Bering Sea coast, or an island within the Bering Sea, meet criteria established by the Governor of Alaska, be a village certified by the Secretary of the Interior pursuant to the Alaska Native Claims Settlement Act, and consist of residents who conduct more than half of their current commercial or subsistence fishing in the Bering Sea or waters surrounding the Aleutian Islands. These villages cannot have previously developed harvesting or processing capacity capable of substantial participation in the Bering Sea fisheries in order to qualify for the program. Currently, the CDQ program allocates 7.5 percent of the total allowable catch in the pollock, halibut and sablefish, crab, and groundfish fisheries to the Community Development Quota Program.

CFEC (Commercial Fisheries Entry Commission): An independent, quasi-judicial regulatory agency responsible for promoting the sustained yield management of Alaska's fishery resources and the economic health and stability of commercial fishing by regulating entry into the fisheries. The CFEC controls the number of permits and qualifying criteria of fishing permits.

CFR (Code of Federal Regulations): A codification of the general and permanent rules published in the Federal Register by the Executive departments and agencies of the Federal Government.

Chamorro: Natives of the Northern Mariana Islands and Guam

charterboat: A boat designed for carrying passengers for hire who are engaged in recreational fishing.

commercial: Fishing where the primary intent of the fishing is to sell, barter or trade the catch.

CNMI (Commonwealth of the Northern Mariana Islands): Also, Northern Mariana Islands, Northern Marianas, and NMI. Commonwealth of the United States in the Western Pacific. Includes the islands of Saipan, Tinian, Rota, and many others in the Marianas Archipelago.

CVFC (Coastal Villages Fishery Cooperative): Community Development Quota Association composed of the villages of Cherfornak, Cheevak, Eek, Goodnews Bay, Hooper Bay, Kipnuk, Konigianok, Kwigillingok, Mekoryuk, Newtok, Nightmute, Platinum, Quinhagak, Scammon Bay, Tooksok Bay, Tuntutuliak, Tununak. The CVFC is the predecessor to the CVRF.

CVRF (Coastal Villages Region Fund): Community Development Quota Association composed of the same villages that composed its predecessor the CVFC. The CVRF was formed in 1997.

DCRA (Department of Community and Regional Affairs): Agency in the State of Alaska responsible for overseeing the management of the Community Development Quota Program.

DFW (Division of Fish and Wildlife): Agency in the Northern Mariana Islands responsible for overseeing fisheries and wildlife management.

EEZ (exclusive economic zone): Zone extending from the shoreline out to 200 nautical miles in which the country owning the shoreline has the exclusive right to conduct certain activities such as fishing. In the United States, the EEZ is split into state waters (typically from the shoreline out to 3 nautical miles) and federal waters (typically from 3 to 200 nautical miles).

Fa'a Samoa: The term means the Samoan Way. This is an all encompassing concept that dictates how Samoans are meant to behave. Fa'a Samoa refers to the obligations that a Samoan owes his or her family, community and church and the individual's sense of Samoan identity.

factory-trawler: (see catcher-processor)

finfish: Fishery species not including crustaceans, cephalopods, or other non-vertebrate species.

fishery management council: Eight regional fishery management councils are mandated in the Magnuson-Stevens Fishery Conservation and Management Act to be responsible for developing fishery management plans for fisheries in federal waters. Councils are composed of voting members from NMFS, state fishery managers, and individuals selected by governors of the coastal states. Nonvoting members include the U.S. Coast Guard, the U.S. Fish and Wildlife Service, and other federal officials. Regional councils exist for the Caribbean, Gulf of Mexico, Mid-Atlantic, New England, North Pacific, Pacific, South Atlantic, and Western Pacific regions.

FMP (fishery management plan): Management plan for fisheries operating in the federal EEZ produced by regional fishery management councils and submitted to the Secretary of Commerce for approval. These plans must meet certain mandatory requirements in the Magnuson-Stevens Fishery Conservation and Management Act before they can be approved or implemented.

Fono: Traditional Samoan village council. The Fono is responsible for adminis-
tering justice within the village and can pass down a wide range of judgments.

GOA (Gulf of Alaska): Region of the EEZ off the shore of Alaska extending
from the southeastern edge of Alaska to the eastern side of the Aleutian Chain.

groundfish: Fish species found on or near the bottom or floor of the ocean (e.g.,
halibut, yellowtail flounder).

Guam: A territory of the United States in the Western Pacific. It is south and
adjacent to the Commonwealth of the Northern Mariana Islands.

handline: A fishing line designed for handlining. Typically, handlines with baited
hooks are dropped and retrieved by hand. They may be trolled.

handlining: A method of fishing in which the fisherman uses one or more
handlines.

highliner: A fisherman who is regarded as having a fishing operation with high
catch and profits.

IFQ (individual fishing quota): Fishery management tool used in the Alaska
halibut and sablefish fisheries and in other fisheries that allocates a certain portion of the TAC to individual vessels or fishermen based on initial qualifying
criteria. This allocation can be transferred or sold. If the IFQ is transferable, it is
sometimes referred to as an individual transferable quota (ITQ).

IP (Imapiqamiut Partnership): A partnership established between Coastal Villages Fishing Cooperative and Golden Age Fisheries. The partnership was dissolved in 1998 due to pressure from the State of Alaska concerning the performance of the partnership.

IPHC (International Pacific Halibut Commission): International management and
advisory body established in 1923 to oversee the management of halibut in the
North Pacific. The member states to the commission include the United States
and Canada. The IPHC is responsible for conducting stock assessments, and
setting the TAC for the stock in a given year.

JV (joint venture): Cooperative arrangements between foreign and U.S. harvesters to prosecute a fishery occurring inside of federal waters. Joint ventures were
used extensively in the late 1970s through the 1980s in the North Pacific fisheries, particularly in the pollock and crab fisheries. Currently, there are no JV
arrangements in the North Pacific EEZ.

longline: Fishing method using a horizontal mainline to which weights and baited hooks are attached at regular intervals. The horizontal mainline is connected to the surface by floats. The mainline can extend from several hundred yards to several miles and may contain several hundred to several thousand baited hooks.

longliner: A vessel specifically designed to catch fish using the longline fishing method.

Manu'a: An island group in the eastern part of American Samoa.

Mau Zone: A fishery zone under the bottomfish FMP of the Western Pacific Council

mothership: Vessel, typically anchored, to which catch is delivered. Typically motherships will process the catch onboard. Most motherships receive catch from several different vessels.

MHI (main Hawaiian islands): The inhabited portion of the Hawaiian islands, composed of Hawaii, Kauai, Oahu, Maui, Molokai, Lanai, Niihau, Kahoolawe, and Kaula.

MSFCMA (Magnuson-Stevens Fishery Conservation and Management Act): Federal legislation responsible for establishing the fishery management councils and the mandatory and discretionary guidelines for federal fishery management plans. This legislation was originally enacted in 1976 as the Fishery Management and Conservation Act; its name was later changed to the Magnuson Fishery Conservation and Management Act, and in 1996 was renamed the Magnuson-Stevens Fishery Conservation and Management Act.

mt (metric ton): 1,000 kilograms (equivalent to 2,206 pounds).

multispecies: A fishery in which more than one species is caught or captured at the same time. In the case of the CDQ program, the multispecies quota includes all commercial species other than crab, halibut, sablefish, and pollock.

NMFS (National Marine Fisheries Service): Federal agency within the Department of Commerce responsible for overseeing fishery science and regulation. NMFS is part of NOAA

NOAA (National Oceanic and Atmospheric Administration): Agency within the Department of Commerce responsible for ocean and coastal management. NMFS is part of NOAA.

NPFMC (North Pacific Fishery Management Council): One of eight regional councils mandated in the Magnuson-Stevens Fishery Conservation and Management Act to develop management plans for fisheries in federal waters off of the state of Alaska. It is comprised of voting members from NMFS, state fishery managers from Alaska, Washington, and Oregon, and individuals selected by the governors of the three states. Alaska has six voting members, Washington has three voting members, and Oregon has one voting member. Nonvoting members include the U.S. Coast Guard, the U.S. Fish and Wildlife Service, and other federal officials.

NSEDC (Norton Sound Economic Development Corporation): Community Development Quota Association composed of the villages of Brevig Mission, Diomede/Inalik, Elim, Gambell, Golovin, Koyuk, Nome, Savoonga, Shaktoolik, St. Michael, Stebbins, Teller, Unalakleet, Wales, White Mountain.

NWHI (Northwestern Hawaiian Islands): All islands in the Hawaiian Island chain Northwest of the Main Hawaiian Islands.

pelagic: The ocean surface, or the open ocean.

PIAFA (Pacific Insular Area Fishing Agreement): Potential fishery agreement between the State of Hawaii, the Commonwealth of the Northern Mariana Islands, American Samoa, the uninhabited U.S. territories in the western Pacific, and foreign fishing nations. Such an agreement, if approved, would allow foreign vessels to fish within the U.S. EEZ by payment of fees in excess of administrative costs. These fees can be used to develop specific fishery-dependent research and development in the region.

purse seine: Fishing for certain species, usually tuna, in which the school of fish is encircled with a large vertical net with a closable bottom. The fish are trapped by closing the bottom of the net.

recreational: Fishing where the primary intent of the fishing is for sport and pleasure and not the sale, barter, or trade of the fish.

setline: A long heavy fishing line to which several hooks are attached in a series.

SSC (scientific and statistical committee): Fishery management advisory body composed of federal, state, and academic scientists that provides scientific advice to a fishery management council.

surimi: A protein paste derived from processing raw fish. Surimi can be combined with flavoring agents and other substances to create marketable foodstuffs (e.g., imitation crab meat).

TAC (total allowable catch): Total catch permitted to be caught from a stock in a given time period, typically a year. In the United States, this limit is determined by fishery management councils in consultation with NMFS and scientific and statistical committees where they are used.

transshipment: To transfer product from one ship to another at-sea for further transport of the product.

trolling: Fishing technique where a lure is attached to a line dragged through the water. This technique is used in fishing for tuna and other pelagic species.

trawling: Fishing technique in which a net is dragged behind the vessel and then retrieved when full of fish. This technique is used extensively in the harvest of pollock, cod, and other species in North Pacific fisheries.

U.S. Bureau of Commercial Fisheries: Fishery regulatory body that preceded the National Marine Fisheries Service, the Magnuson-Stevens Fishery Conservation and Management Act, and the establishment of the fishery management councils.

WPRFMC (Western Pacific Regional Fishery Management Council): One of eight regional councils mandated in the Magnuson-Stevens Fishery Conservation and Management Act to develop management plans for fisheries in federal waters off of the state of Hawaii, Guam, the Commonwealth of the Northern Mariana Islands, American Samoa, and uninhabited territories in the Western Pacific. It is comprised of voting members from NMFS, state fishery managers from Hawaii, territorial representatives from the Commonwealth of the Northern Mariana Islands, Guam and American Samoa, and individuals selected by the governors of the four states. Hawaii has 5 voting members, the Northern Mariana Islands has 2 voting members, Guam has 2 voting members, and American Samoa has 3 voting members. Nonvoting members include the U.S. Coast Guard, the U.S. Fish and Wildlife Service, and other federal officials.

YDFA (Yukon Delta Fisheries Development Association): Community Development Quota Association composed of the villages of Alakanuk, Emmonak, Kotlik, and Sheldon Point.